# 100 Years of Radio in South Africa, Volume 2

"Dr Sisanda Nkoala and Prof Gilbert Motsaathebe have managed to historicize, contextualize, and theorize the evolution and maturation of radio as a medium in South Africa. Although the book focuses on the last 100 years of radio broadcasting in South Africa, most of the themes, nuggets and case studies that it touches on can easily be used to explain what is happening in similar global South contexts. This is a lucid scholarly work which foregrounds the resilience, relevance, permanency, and adaptability of radio as a medium of communication. It is a historical treatise of significant proportions."

—Admire Mare, *Associate Professor, Department of Communication and Media, University of Johannesburg*

Sisanda Nkoala • Gilbert Motsaathebe
Editors

# 100 Years of Radio in South Africa, Volume 2

## Community Radio, Digital Radio and the Future of Radio in South Africa

palgrave
macmillan

*Editors*
Sisanda Nkoala
Cape Peninsula University of Technology
Cape Town, South Africa

Gilbert Motsaathebe
ILMA, North-West University
Mahikeng, South Africa

ISBN 978-3-031-40705-5 ISBN 978-3-031-40706-2 (eBook)
https://doi.org/10.1007/978-3-031-40706-2



This Palgrave Macmillan imprint is published by the registered company Springer Nature Switzerland AG.
The registered company address is: Gewerbestrasse 11, 6330 Cham, Switzerland

Paper in this product is recyclable.

*In commemorating a century of radio in South Africa, we dedicate this book to the old and future generations of radio practitioners. As we reflect on the remarkable evolution of community and online media, we look ahead with hope and anticipation for the innovative storytellers, broadcasters, and visionaries yet to come.*

*We also dedicate this work to the radio broadcasting students and academics at all institutions. With the knowledge contained within these pages, we hope you find valuable insights, historical context, and the encouragement to continue pushing the boundaries of radio, both over the air waves and in the digital realm.*

*Abazithobileyo*

*Sisanda Nkoala and Gilbert Motsaathebe*

# Preface

Both community radio and online media have played a significant role in the South African media landscape in recent years. Community radio has been particularly important in providing a voice to marginalised communities and promoting social inclusion since it was a recognised tier of broadcasting in South Africa in the 1990s. On the other hand, online media has enabled people to access radio content from all over the world, regardless of their location, amid the growth in internet penetration and affordability in this context. As the year 2023 marks 100 years of radio in South Africa, this edited book, the second of a 2-volume series on 100 years of radio in South Africa, focuses on the two recent additions to South African broadcasting: community radio and online media. Community radio is a type of radio broadcasting owned and controlled by the community it serves. It is typically non-profit and independent and aims to provide a platform for local voices and perspectives. Online media, on the other hand, refers to audio broadcasting that is distributed over the internet. This includes internet radio stations, podcasts, and other audio content that can be accessed online.

While these platforms may be relatively new compared to public and commercial broadcasters, they constitute the majority of radio stations in terms of sheer numbers. They offer a unique space for alternative voices, enabling grassroots organisations, activists, artists, and community members to share their perspectives and stories and address pertinent social, cultural, or political issues. Additionally, community and online radio have been quick to embrace technological advancements. Community stations are increasingly adopting digital technologies for broadcasting and production, while online platforms leverage the internet and streaming technologies to reach global audiences. What makes them significant is their shared commitment to localised content, adaptability, accessibility, and their role as alternative platforms for broadcasting and content consumption beyond traditional mainstream media.

This volume brings together a variety of perspectives on the role of community radio and online media in South Africa. The chapters in this volume are written by academics, practitioners, and activists who have a deep understanding of these media. They explore the history, development, and impact of community radio and online media in South Africa. They also discuss the challenges and opportunities these media face in the future. One of the key contributions the volume makes to the scholarly literature on radio in South Africa is the inclusion of a timeline and historgraphy of the South African Community and Online Radio from 1988 to 2014.

The volume begins by spotlighting community radio, which has burgeoned with over 200 stations nationwide and new ones being established regularly. It is divided into two parts. The first part focuses on community radio and explores the history, development, and impact of community radio in South Africa. The chapters in this part are written by academics, practitioners, and activists involved in the community radio movement. The second part is on online media and examines the rise of online audio media in South Africa and its impact on the radio landscape. The chapters in this part are written by academics who have been studying the development of online media.

While public and commercial broadcasters have played pivotal roles, community radio and online platforms have emerged as dominant players in numbers and alternative perspectives. They offer spaces for grassroots organisations, activists, artists, and community members to express themselves and address social, cultural, and political issues. Technological advancements have further enhanced their reach and impact. Through thoroughly exploring these emerging platforms, this volume aims to provide insights into the direction the sector may take in the next century.

Cape Town, South Africa — Sisanda Nkoala
Mahikeng, South Africa — Gilbert Motsaathebe

# A Timeline and Histography of South African Community and Online Radio 1988–2014

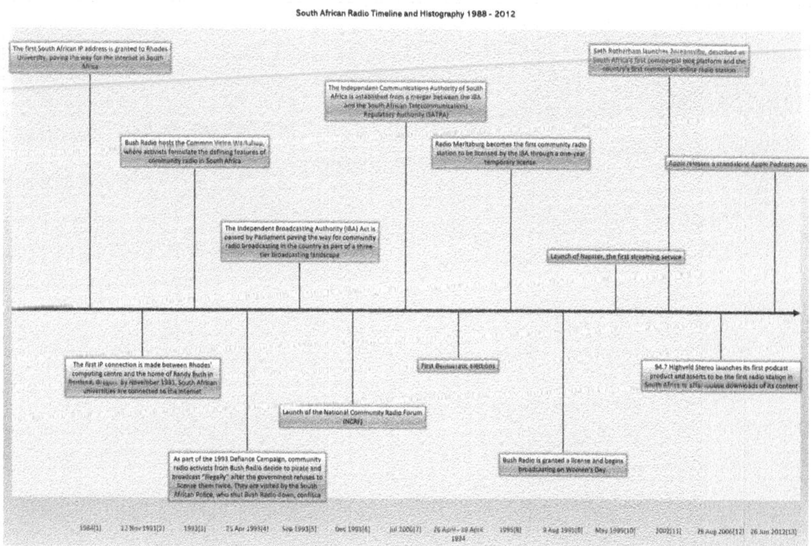


**Fig. 1** A timeline of key developments in South African community and online radio

Cape Peninsula University of Technology
Cape Town, South Africa

Sisanda Nkoala

# A Timeline and Histography of South African Community and Online Radio 1988–2014 (Table)

| *Date* | *Event* |
|---|---|
| 1988[a] | The first South African IP address is granted to Rhodes University paving the way for the internet in South Africa |
| 12 Nov 1991[b] | First IP connection is made between Rhodes' computing centre and the home of Randy Bush in Portland, Oregon. By November 1991, South African universities are connected to the internet. |
| 1993[c] | Bush Radio hosts the Common Vision Workshop where activists formulate the defining features of community radio in South Africa |
| 25 Apr 1993[d] | As part of the 1993 Defiance Campaign, community radio activists from Bush radio decide to pirate and broadcast "illegally" after the government refuses to license them twice. They are visited by the South African Police, who shut Bush Radio down, confiscate its equipment and arrest some of the activists. This is viewed as Bush Radio's first broadcast, even though it was not licensed at this stage |
| Sep 1993[e] | Independent Broadcasting Authority (IBA) Act passed by Parliament paving the way for community radio broadcasting in the country, as part of a three-tier broadcasting landscape |
| Dec 1993[f] | Launch of the National Community Radio Forum (NCRF) |
| Jul 2000[g] | Independent Communications Authority of South Africa (ICASA) is established from a merger between the IBA and the South African Telecommunications Regulatory Authority |
| 26–29 Apr 1994 | First democratic elections |

(*continued*)

| *Date* | *Event* |
|---|---|
| 1995[h] | Radio Maritzburg becomes the first community radio station to be licensed by the IBA through a one-year temporary licence |
| 9 Aug 1995[i] | Bush Radio is granted a licence and begins broadcasting on Women's Day |
| May 1999[j] | Launch of Napster, the first streaming service |
| 2002[k] | Seth Rotherham, founder and CEO of 2oceansvibe, described as South Africa's first commercial blog platform and the country's first commercial online radio station |
| 28 Aug 2006[l] | 94.7 Highveld Stereo launches its first podcast product and asserts that it is the first radio station in South Africa to offer mobile downloads of its content |
| 26 Jun 2012[m] | Apple releases a standalone Apple Podcasts app |
| 1 May 2014[n] | Launch of Cliffcentral, an online content hub that specialises in podcasting and online radio started by former 5FM radio presenter Gareth Cliff |
| 29 Aug 2016[o] | Thabo "TBo Touch" Molefe partners with Cliffcentral to launch Touch Central, an online platform which describe itself as "South Africa's first ever 24-hour online music channel" |

[a]http://archive.hmvh.net/txtfiles/interbbs/SAInternetHistory.pdf

[b]http://archive.hmvh.net/txtfiles/interbbs/SAInternetHistory.pdf

[c] Lumko Mtimde (2023)

[d] Lumko Mtimde (2023, this volume)

[e]https://www.icasa.org.za/uploads/files/Independent-Broadcasting-Authority-Act-153-of-1993.pdf

[f]https://www.comminit.com/media-development-africa/content/national-community-radio-forum-ncrf-1

[g]https://www.icasa.org.za/pages/our-mandate#:~:text=ICASA%20was%20established%20in%20July,the%20Postal%20Regulator%20into%20ICASA

[h]https://www.dut.ac.za/radio-expert-addresses-journalism-students/

[i]https://web.archive.org/web/20130410022707/http://www.transformationsjournal.org/journal/issue_10/article_05.shtml

[j]https://www.michigandaily.com/music/brief-history-steaming-services/

[k]https://www.2oceansvibe.com/about/

[l]https://www.mediaupdate.co.za/media/2624/947-highveld-stereo-launches-radio-content-on-mobile-phone

[m]https://www.wired.com/2012/06/apple-podcasts-app/

[n]https://cliffcentral.com/about-us

[o]https://www.news24.com/channel/5-controversial-gareth-cliff-moments-that-had-us-on-the-edge-of-our-seats-20161201

Cape Town, South Africa Sisanda Nkoala

# Acknowledgements

We are deeply grateful to the individuals and institutions that made this edited volume possible. This book is a testament to the collective effort of a diverse and talented group of people who shared their insights and worked tirelessly to bring it to fruition.
We would like to especially thank the following contributors:

- All of the authors who contributed their valuable research and expertise. Their chapters are essential to the book's comprehensiveness and rigour.
- The peer reviewers who provided constructive feedback and rigorous assessments. Their insights helped to improve the quality of the book.
- Our critical readers and the esteemed scholars who wrote reflections of endorsement for this volume.
- The publisher, Palgrave Macmillan, for believing in this project and assisting to bring it to fruition.
- The Cape Peninsula University of Technology's (CPUT) Faculty of Informatics and Design that provided funding and resources. The support that made it possible for us to bring this book to fruition was invaluable.
- The South African Broadcasting Corporation's Western Cape branch and South African National Editors' Forum for having assisted with hosting a seminar for CPUT students based on some of the contents of this book.

- Last, but not least, the family and friends who provided support, encouragement, and understanding. Their love and support helped us to persevere through the challenges of editing this book.

We are deeply grateful to each of you for your invaluable contributions. This book would not have been possible, otherwise.

Sisanda Nkoala and Gilbert Motsaathebe

# Contents

# Notes on Contributors

**Joseph Adebayo** is a lecturer in the Media Studies Department at Birmingham City University, Birmingham, UK. Adebayo holds a PhD from Durban University of Technology, where his thesis focused on peace journalism. He is the author of the book *Reporting African Elections: Towards a Peace Journalism Approach* (2018). His research interests include peace journalism and society, elections, good governance, and democracy in Africa.

**Kealeboga Aiseng** is a lecturer in the Journalism and Media Studies Department at Rhodes University, Makhanda, South Africa. Aiseng holds a PhD in African Languages and Linguistics from the University of the Witwatersrand. Aiseng's research interests are sociolinguistics, language policy, African popular culture, new media and film studies. He has published articles in local and international journals. He also serves on numerous journal editorial boards.

**John Bulani** is a lecturer and the Work Integrated Learning coordinator in the Media Department at Cape Peninsula University of Technology, Cape Town South Africa. Bulani is a media scholar and practitioner. He holds an MTech in Public Relations Management from the Cape Peninsula University of Technology. He serves as an editorial coordinator for local community newspaper *Khwezi Times News*. His research interests include community media, service learning and citizen journalism.

**Freeborn Isaac Kibombwe** is a lecturer in Theology, Media Studies and Public Relations at Cavendish University of Zambia, Lusaka, Zambia, and a

Catholic Priest from Zambia. Kibombwe holds a DMin (Doctor of Ministry) in Pastoral Theology and Communications. His work is in the religious and social communication interdisciplinary field. He has published in the area of theology and communications media. He is an active member of the Media Institute of Southern Africa (MISA-Zambia Chapter).

**Carolyne M. Lunga** is a senior lecturer in Journalism and Media Studies at University of Portsmouth, Portsmouth, UK. Lunga holds a PhD in Journalism. Her work is on investigative journalism, digital journalism, global media practices and media ethics. She has published on investigative and digital journalism. She is an editorial board member of Journalism Practice.

**Trust Matsilele** is a senior lecturer in Journalism at Birmingham City University, UK. Matsilele holds a PhD from the Department of Communications at the University of Johannesburg. He researches the changing journalism ecologies in Africa. Matsilele has published in international journals, including *Media, Culture & Society*, *Digital Journalism* and the *Journal of Communication Inquiry*.

**Quatro Mgogo** is a senior lecturer in the Faculty of Humanities at the North-West University, South Africa. Mgogo hold a doctoral degree in Media and Communication from the University of Fort Hare, South Africa. Mgogo is an alumnus and a doctoral mentor of the National Institute for Humanities and Social Sciences. His research interests are in media representation, media and violence, media and indigenous language, and intercultural communication.

**Tshepang Molale** is a senior lecturer based at the International Association for Minority Language Media Research Unit. Molale holds a PhD in Communication Studies. He has published several papers in accredited journals and has contributed with book chapters within the fields of Development Communication, Journalism and Media Studies.

**Gilbert Motsaathebe** is a full professor attached to the Indigenous Language Media in Africa (ILMA) research entity at the North-West University, Mahikeng, South Africa. Motsaathebe hold a PhD in Film and Media Studies from the University of the Witwatersrand. He was, until recently, Editor-in-Chief of *Communicare: Journal for Communication Studies in Africa*. He serves on the Panel of Judges for South Africa's premier competition of journalists, the Vodacom Journalist of the Year. He is a National Research Foundation-rated researcher with numerous peer-reviewed publications including ten books in different genres.

**Phillip Mpofu** is an Extraordinary Researcher in the Indigenous Language Media in Africa (ILMA) research entity. His research interests are at the intersection of language and media studies. He holds DLitt et Phil; BA; MA African Languages and Culture; PG Diploma in Media and Society Studies.

**Lumko Mtimde** is an award-winning Activist, Former Special Advisor (Minister in the Presidency), retired public servant, former CEO of several public entities, and former Councillor of key ICT regulators in South Africa. Mtimde holds a BSc and postgrad diploma (Honours). He has published widely in South African media and abroad with his work mentioned in Academia.edu. He serves on several boards and councils of higher education Institutions in South Africa.

**Rosemary K. Musvipwa** is a lecturer in Journalism and Mass Communication at University of Eswatini, Kwaluseni, Eswatini. Musvipwa holds an MSc in Journalism and Media Studies. Her focus is media studies, health communication and sustainable development. She is a PhD candidate in Culture, Communication and Media Studies at the University of KwaZulu-Natal.

**Ofentse Ngake** is a lecturer in Journalism, Communication and Media Studies at North-West University. He is an emerging scholar with research interests in media studies and social change, journalism and indigenous language media. He has co-supervised a number of honours studies and presented at research conferences.

**Sisanda Nkoala** is a senior lecturer in Media Studies at Cape Peninsula University of Technology, South Africa. Nkoala holds a PhD in Rhetoric Studies. Her work focuses on the rhetoric of and by South African media ecologies. She has published widely on South African media, multilingualism and rhetoric. She is a public representative on the South African Press Council, among other roles within the South African media landscape.

**Abiodun Salawu** is a professor of Journalism, Communication and Media Studies at the North-West University, Mahikeng, South Africa, and Director of the research entity Indigenous Language Media in Africa (ILMA). He has taught and researched journalism, media and communication for close to three decades. He has scholarly publications in academic journals and books. He has also edited/co-edited eleven books and authored one.

# List of Figures

# List of Tables

CHAPTER 1

# Community and Online Radio as Emerging Frontiers of South African Audio Broadcasting

*Sisanda Nkoala and Gilbert Motsaathebe*

## About This Volume

South Africa has three tiers of broadcasting, namely public, commercial, and community broadcasting. Accordingly, the country essentially has three forms of radio stations. In recent years, however, the internet and its streaming technology have introduced other forms of audio broadcasting such as internet radio and more lately podcasts. Therefore, this volume (Volume 2) builds on 100 Years of Radio in South Africa 1923 - 2023 Volume 1 by focusing on community radio and online media, the recent additions to South African audio broadcasting.

S. Nkoala (✉)
Cape Peninsula University of Technology, Cape Town, South Africa

G. Motsaathebe
ILMA, North-West University, Mahikeng, South Africa



S. Nkoala, G. Motsaathebe (eds.), *100 Years of Radio in South Africa, Volume 2*, https://doi.org/10.1007/978-3-031-40706-2_1

In our view, the history of radio in South Africa would be incomplete without consideration of the role of community radio and online radio. These platforms may be relatively new, but compared to the public broadcasters and commercial stations, they make up the lion's share of radio stations in terms of numbers. They provide space for alternative voices and perspectives, and offer a platform for grassroots organisations, activists, artists, and community members to express their perspectives, share stories, and address social, cultural, or political issues. Community and online radio have also benefited particularly from technological advancements in recent years. The former are increasingly embracing digital technologies for broadcasting and production, while the latter leverage the internet and streaming technologies to reach global audiences. It is our view that because of the similarities in their orientation towards localised content and engagement, flexibility, and adaptability, as well as their accessibility as alternative platforms for broadcasting and accessing content outside of traditional mainstream media, they are well placed to be considered together in this volume.

Beginning with community radio, there are over 200 community radio stations across the country, with new ones being established every year. For example, the Independent Communications Authority of South Africa recently called for applications for Community Radio Broadcasting Services. This type of radio plays a unique role as it has the potential to focus on local issues affecting specific target communities compared to mainstream stations. Similarly, a cursory review shows that online-based radios continue to evolve at an exponential rate. Internet-based radio listenership is driven by the fact that online media platforms and tools have become part and parcel of modern society. The 2022 census figures show that 79% of South Africans own a device that has some measure of internet connectivity, allowing them to access online platforms (StatsSA 2023). The online platforms have also provided opportunities for broadcasters in that space to distribute their content simultaneously via multiple platforms available on the internet through technological convergence. For instance, some online radios and podcasts often have simulcasts via YouTube channels and even mainstream television channels, effectively transforming this form of radio from being strictly an audio medium to an audio-visual one.

Therefore, from this vantage point, it is clear that more scoping work needs to be done on the prevalence of online audio platforms, such as online radio stations and podcasts. This volume thus attempts to fill this lacuna, through the variety of chapters that provide depth on the operation of community radios and online radio platforms in the country. Such a scholarly examination, coupled with practitioners' perspectives, is well placed in a text that considers the role of radio over the last 100 years in South Africa because it also assists in forecasting where the sector may be headed in the next century.

## A Voice from the Grassroots: Community Radio in South Africa

The volume starts off with a chapter from one of the founders of the community radio sector in South Africa, Lumko Mtimde. In his chapter, Mtimde provides a detailed account of how the community radio movement started in South Africa and recalls that established broadcasters of the pre-democracy era were unconvinced by the idea of setting up community radio stations in the country. The chapter details how community radio activists, under the banner Freedom of the Airwaves campaign, were ridiculed and socially castigated when they called for the establishment and licensing of community radio. Today, one cannot imagine a South African radio sector without the over 200 community stations spread across the country.

The four chapters following Mtimde's comprehensive overview of community radio consider issues such as the role of this sector in language revitalisation and preservation of African culture, economic empowerment of youth, and religion. Using various case studies, these chapters contextualise the issues facing the selected community radio stations they profile. In his chapter, Kealeboga Aiseng considers Valtaar FM, a Setswana-language community radio station, to examine the role played by indigenous language community media in revitalising indigenous languages. His study draws influence from indigenous radio stations as a catalyst for protecting and preserving languages from total extinction. John Bulani, Joseph Adebayo, and Trust Matsilele examine the role community radio plays in fostering the economic empowerment of youth in the Cape Flats, using Bush Radio as a case study. This chapter begins by outlining the role of community radio in social inclusion and considers the establishment of the sector in light of the 100 years of the existence of radio in this context. It then expounds on why Bush Radio, as the oldest community radio station in South Africa, is an interesting case study to consider when one examines the place of community radio in youth empowerment.

Father Freeborn Isaac Kibombwe who established Oblate Radio Liseli (Light) in Zambia tells the story of Radio Veritas, the only Catholic radio station in South Africa. He tells the story from his perspective as someone with a close eye on how the process unfolded, as well as someone who has had experience setting up a Catholic radio station in another context. Radio Veritas' story is one of resilience, hard work and, above all, the faith and optimism of the founder, Father Emil Blaser, and a few committed and dedicated staff members who never gave up trying to get the station on the air and keeping it there over the past 23 years since it was launched in 1999.

Tshepang Molale and Quatro Mgogo's chapter aptly titled "More Than Just Edutainment: Indigenous South African Community Radio Music Programmes as Preservers of African Traditional Cultures" reports on research conducted on two indigenous language community radio stations in South Africa, namely, Inkonjane Community Radio in the Eastern Cape, which broadcasts in Isimpondo (a sub-language of IsiXhosa), and Mahikeng FM, which broadcasts in Setswana in the North West province. It argues that community media, particularly indigenous language music programmes, are integral in that they can help facilitate local culture and everyday life. Against the backdrop of 100 years of radio broadcasting in South Africa, this chapter is aimed at transcending the age-old motifs that black radio only serves to educate, inform, and entertain audiences and instead considers radio stations aimed at marginalised community development in their approach in building communities that embrace their cultures through traditional music, dance, and poetry.

## Online Engagements That Support On-Air Interactions: The Role of Digital and Social Media in South African Radio

In Chap. 6, Ofentse Ngake, Abiodun Salawu, and Phillip Mpofu examine the use of the WhatsApp voice note service on Motsweding FM as a case study that demonstrates how digital technology promotes the use of indigenous languages in the digital public sphere of radio. They draw on convergence theory and the diffusion of innovation theory to argue that some presenters on these kinds of stations use the WhatsApp voice note service generally for audience participation on the radio programme. Their study further finds that promoting indigenous languages on the radio's digital public sphere through the WhatsApp voice note service is practical.

Chapters 7 and 8 provide a cogent link between community radio and online radio, in their consideration of how terrestrial radio stations use online platforms to support on-air interactions. Carolyne M. Lunga and Rosemary Musvipwa's chapter, "Radio production in the Digital Era: Lessons From South Africa", focuses on two commercial stations, 947 and Kaya FM, and two stations owned by the South African Broadcasting Corporation (SABC), UKhozi FM and 5FM, to examine how selected South African radio stations are navigating convergence and digitisation within the radio industry. By drawing on network theory and adopting a textual analysis of the stations' digital platforms, they argue that their online presence shows that they have adapted to the digital era and take advantage of social networking sites to connect more with their audiences, expand their listenership base, and form local and global networks. These interconnections are thus impacting the content production, dissemination, and consumption processes.

Compared to the more technologically advanced contexts in the global north, the uptake of social media by South African radio stations, and the ability to incorporate these technologies as part of programming strategies with the potential to create counter-publics, was slow (Bosch, 2014). However, as Lunga and Musvipwa, and Ngake, Salawu, and Mpofu, referenced above, note in this volume, online engagement facilitated by social media has become such an integral part of how South African radio stations interact with listeners online that 100 years into the lifespan of radio in this context, it has become difficult to imagine the medium before this. The final section of this volume considers platforms that are exclusively digital in the form of online radio stations and podcasts.

## Online Radio and Podcasts: An Emerging Alternative Digital Public Sphere

The exact date of birth of online radio and podcasting in South Africa, as we know it, is difficult to ascertain. No doubt, following the introduction of the internet in the early 1990s, those with access experimented with various forms of self-expression using the technology, as the amateurs who had first played around with radio in the 1920s did. One entity, however, views itself as a pioneer in professionalising and commercialising the use of the internet for broadcasting—2oceansvibe. Founded by marketing graduate Seth Rotherham in 2002, 2oceansvibe was initially a platform used by Rotherham to post photographs of his adventures (2oceansvibe Media, 2022). It grew to become a media and news entity that features the 2oceansvibe radio platform that live streams 24 hours a day.

Another milestone in online radio and podcasting was the launch of Cliffcentral, an online content hub that specialises in podcasting and online radio founded by former 5FM presenter Gareth Cliff after he left the station (Cliffcentral.com, 2022). Describing the platform, Cliff dubbed it an "'un-radio' venture that aims to merge online broadcasting with social media and instant messaging platforms" (Macguire, 2014). In media interviews on what gave rise to him starting the platform, Cliff would highlight the lack of regulation on internet-based broadcasting as one of the features that motivated him to move from being on-air to being online (Macguire, 2014). During his ten-year career at the South African Broadcasting Corporation (SABC) Cliff built a reputation as a shock jock. In 2004, for example, the Broadcasting Complaints Commission of South Africa found that he had contravened its code of protecting religious freedom, following a mock interview he pretended to conduct with "Jesus" on his show (BizCommunity, 2004). Cliff was also motivated by a desire to connect directly with his listeners using the online radio platform and associated social media sites, such as Facebook and Twitter.

For Cliff, and many other South African podcasters and online radio hosts, these platforms present an alternative digital public sphere that, to a larger degree than not, is not subjected to government regulation. For this former radio broadcaster, the prospect of more direct engagement with his audience and less red tape in terms of content presented by online media has undertones of some of the features that community radio activists touted as a reason to introduce the sector in the early 1990s. With the introduction of community broadcasting, the activists saw untapped possibilities in using existing technology to engage in new ways. Online radio, on the other hand, presented a new technology and a new way of broadcasting by reimagining radio production practices.

Questions of impact have, however, plagued the sector, with queries raised around the reach of these stations and the audience figures they report which some claim have been inflated (Muller, 2014). Another issue pertains to access and the high cost of data, which means that many South Africans who might consider starting online radio stations will struggle to attract listeners outside the Living Standard Means 7—10 categories. This was one of the issues that then-former Metro FM presenter Thabo "TBo Touch" Molefe said he wanted to champion when he joined forces with Cliff, in 2016, to launch Touch Central, "South Africa's first-ever 24-hour online music channel" (News24, 2016). In September 2016, Molefe gave South African mobile networks an ultimatum to change in 30 days or he would ensure that everyone moved to a network "that cares about its customers". This was supposedly part of the #DataMustFall campaign. By November of that year, it emerged that the campaign had not genuinely been about advocating for consumers, but was rather a publicity stunt, when Molefe started advertising products for one of the operators, MTN (Vermuelen, 2016).

Despite these challenges, some studies predict that the monthly listenership of podcasts could rise to 19 million by 2024 (PwC, 2022), with speculation that this is where growth in terms of audience lies. In her chapter on indigenous language podcasts in South Africa, Sisanda Nkoala highlights how this growth may bode well for indigenous language media. She argues that even though terrestrial radio continues to be the most dominant media platform for black South Africans, the rise in indigenous language podcasts marks a significant milestone in the South African media landscape for speakers of these languages, who have previously been afforded limited agency in the sector. The place of podcasts in South Africa must be understood from the rich 100 years of radio in this context, where oral media has been a central platform for developing linguistic and cultural identities.

In his chapter, Gilbert Motsaathebe uses the thriving Podcast and Chill with MacG as an entry point to explore trends and patterns of podcast usage in South Africa from an audience reception perspective. He argues that the success of this podcast is attributable to several factors, which mainly include the type of content, the personality of the host, and the ability to share the content on platforms such as YouTube and Twitter, which enable users to engage with that content effectively. Ultimately, the chapter draws on users' comments to evaluate how they feel about the podcast series and further draws on content analysis of the episodes to delineate thematic areas that characterise the content of this popular podcast series.

At its core, the aim of this volume is to contribute to understanding the evolution of South African radio with a focus on the new frontiers being charted by community and online radio, with the hope of providing insights into the direction the sector may take in the next century. The volume serves as a crucial addition to the scholarship on radio in South Africa. Its focus is not on providing a comprehensive historical narrative, although that comes through; rather the aim is to understand the community and online audio broadcasting case studies considered against the backdrop of a century of audio broadcasting in South Africa.

## References


2Oceansvibe Media. (2022). *2Oceansvibe Media*. http://2oceansvibemedia.com/

BizCommunity. (2004). *Cliff Caught in BCCSA Crossfire*. https://www.bizcommunity.com/Article/196/59/3481.html

Bosch, T. (2014). Social Media and Community Radio Journalism in South Africa. *Digital Journalism, 2*(1), 29–43.

Cliffcentral.com. (2022). *About Us*. https://cliffcentral.com/about-us/

Macguire, E. (2014). *South Africa Shock Jock's 'uncensored, un-radio' Station on WeChat*. CNN Business. https://edition.cnn.com/2014/12/05/business/wechat-cliff-central-radio-south-africa/index.html

Muller, R. (2014, July 4). Online Radio Alive and Kicking in SA. https://mybroadband.co.za/news/internet/54123-online-radio-alive-and-kicking-in-sa.html

News24. (2016). *Tbo Touch Joins Forces with Cliff Central to Launch Touch Central*. https://www.news24.com/Channel/tbo-touch-joins-forces-with-cliff-central-to-launch-touch-central-20160729

PwC. (2022). *PwC Africa Entertainment & Media Outlook' 2022–2026*. https://www.pwc.com/ke/en/publications/entertainment-and-media-outlook.html

StatsSA. (2023). 2022 Census. https://census.statssa.gov.za/#/

Vermuelen, J. (2016). *#DataMustFall Was a Sham and Touch Central Sold Out*. https://mybroadband.co.za/news/columns/189874-datamustfall-was-a-sham-and-touch-central-sold-out.html

CHAPTER 2

# Creating an Enabling Policy Environment for Democratic Participation Through Community Radio: Perspectives of a Practitioner

*Lumko Mtimde*

## Introduction

The story of the establishment and growth of community radio in South Africa is fundamentally a story about freeing the airwaves. The pre-1990 airwaves were characterised by media entities set up to advance apartheid propaganda. The now public service broadcaster, the South African Broadcasting Corporation (SABC), was then a mouthpiece for the National Party. At the same time, the Transkei-Venda-Bophuthatswana-Ciskei homeland services were viewed by many as puppets of the state. Further, there were only two commercial audio broadcasters, namely

L. Mtimde (✉)
Media and ICT Policy & Regulation Expert, Eastern Cape, South Africa



S. Nkoala, G. Motsaathebe (eds.), *100 Years of Radio in South Africa, Volume 2*, https://doi.org/10.1007/978-3-031-40706-2_2

Capital Radio 604 and Radio 702, and one "Afrikaner broederbond"[1] owned private pay television service, MNET. It would take mobilisation and struggle by an alliance of anti-apartheid groups under the auspices of the Mass Democratic Movement (MDM) and the United Democratic Front (UDF) to bring together young people from organisations such as the South African National Students Congress (SANSCO), South African Youth Congress , and South African Students Press Union (SASPU) to drive the transformation of a campus radio station into a campus-based community radio station and, subsequently, the birth of the first community radio station, Bush Radio in Cape Town. These campaigns include the Campaign for Independent Broadcasting (CIB) and SASPU's Free-The-Airwaves campaign led in partnership between the campus radio, University of the Western Cape (UWC) Radio, and the then non-governmental organisation Cassette Education Trust (CASET). Student activists like me became community radio activists through engagements with these and other like-minded organisations.

The story I tell in this chapter is from my humble perspective as a rural boy from the dusty village of Mvenyane in the Alfred Nzo District, Matatiele, Eastern Cape, who would one day grow up to be involved in starting the first community radio in South Africa, Bush Radio, and one of the people responsible for the country's broadcasting policy and an advisor in the presidency. This chapter's perspectives are based not only on what I have read and learnt in theory about the sector but also on what I have lived out personally and practically.

I begin the story by locating community radio in the broader history of broadcasting in the country. Whereas the broadcasting history in South Africa is well documented in many books,[2] and historically dominated by the SABC, including the phenomenal strides in the first 27 years of the country's democracy, reflections in this chapter are contextualised from a personal experience as one of the founding members of the community media sector in South Africa. The SABC is one of the oldest broadcasters in the world, alongside the Canadian Broadcasting Corporation and the Australian Broadcasting Corporation. It was a state broadcaster until it was transformed into a public broadcaster in 1993 through the enactment of the IBA Act 153 of 1993, subsequently guided

[1] The Afrikaner Broederbond was a secret, exclusive Afrikaner Calvinist male organisation in South Africa dedicated to the advancement of Afrikaner interests.

[2] Broadcasting in South Africa – National Association of Broadcasters (nab.org.za).

by the Independent Broadcasting Authority (IBA) Triple Inquiry Report and effectively relicensed in 1999. SABC's first radio station turned 84 in 2022. This then-state broadcaster was launched on 1 August 1936 in Durban with the English service Programme "A" as its first radio channel. In 1985, it was renamed Radio South Africa and, ten years later, rebranded SA FM[3] with "the news and information leader" as its motto. The SABC has 19 radio stations and is one of the country's biggest state-owned enterprises.

Comparably, community radio is new in South Africa. It was only allowed after the 1994 democratic elections. There is, therefore, limited background on community radio in the country before 1994. South Africa's first community radio station, Bush Radio, was established in the early 1990s though it was only licensed years later. It operated as a pirate radio station, and though it got shut down by the apartheid regime, it was joined by Radio Zibonele, which pirated until the IBA formally licensed it. Radio Maritzburg was the first community radio station to be licensed by the IBA through a one-year temporary licence in 1995.

Established broadcasters of the pre-democracy era were unconvinced by the idea of setting up community radio stations in our country. When community radio activists, under the Freedom of the Airwaves campaign, called for establishing and licensing community radio and television, we were characterised as insane and socially castigated because we were the youth. Guided by dialectics, the interlinkage of phenomena and human responses, my appeal is that we analyse these historical developments from our accurate times of existence, and social and political realities. The object of change was evident in our minds as young people, and it was to ensure communities own their means of communication so that, as a society, we could have a dialogue with two-way communication to enhance people's democracy. This, to us, was our theory of change; it was what was to make the needle move and make an impact in the lives of our communities.

The struggle and campaigns to free the airwaves embarked on by media diversity and community radio activists, SASPU, and MDM structures led to the enactment of the Independent Broadcasting Authority Act in December 1993 and recognition of the principle of the independent regulation of broadcasting through the Interim Constitution of 1993, the birth of IBA, now the Independent Communications Authority of South

[3] One of the South African Broadcasting Corporation (SABC) radio stations broadcasting in English.

Africa (ICASA), the Media Development and Diversity Agency (MDDA), and the licensing of more than 200 community radio stations throughout every district municipality of South Africa. The history compass tells us the success story of community radio in this context, its impact on society, and the conversations at the grassroots level towards participatory democracy. Against all odds, community radio and media diversity activists rose above the noise and the crowd. As of 2022, there were more than 200 licensed community radio stations and four licensed "community television" stations.

## Background

Community radio has changed the South African broadcasting landscape. Communication is no longer the realm of the elite. Therefore, community radio in South Africa can be best described as the People's Radio. In general, community radio and media are participatory platforms and mediums of communication. These entities are owned and controlled by the communities they serve. Theoretically, one can locate community radio in the democratic participant theory. The Electronic Communications Act, No. 36 of 2005, defines "community" as including a geographically founded community or any group of persons or public sector with a specific, ascertainable common interest. It further defines "community broadcasting service" as a broadcasting service that

(a) is fully controlled by a non-profit entity and carried on for non-profit purposes;
(b) serves a particular community;
(c) encourages members of the community served by it or persons associated with or promoting the interests of such community to participate in the selection and provision of programmes to be broadcast in the course of such broadcasting service. Furthermore,
(d) may be funded by donations, grants, sponsorships, advertising, membership fees, or any combination of those mentioned above[4]

In the event of making a profit, a community broadcasting service licensee must utilise or invest the funds in the community served for community development. A community broadcasting service licensee must

[4] Electronic Communications Act No 36 of 2005.

submit a report annually with supporting documents to ICASA detailing how it has utilised or invested the profit within the financial year. The submissions include submitting proof of community participation at every annual general meeting. ICASA regulations require that a community broadcasting licensee ensure that ownership of the community broadcasting licensee remains with the community being served, involves the community in the management of the community broadcasting licensee, and establishes programming committees to enable community members to participate in the selection and provision of programmes. The programming committees must be representative of different interest groups within the community served, such as youth, women, or people with disabilities.

This definition and these requirements were informed by community radio activists' inputs, returning from a workshop hosted by Bush Radio in 1993 called the "Common Vision Workshop". It was held at Chapmans' Peak in Cape Town, with submissions made by the National Community Radio Forum (NCRF) to IBA and the Ministry of Communications. Further, the community media sector led by Bush Radio, SASPU, subsequently the NCRF, and the then National Community Media Forum (NCMF) actively submitted policy and regulatory proposals to the Viljoen Commission, the Independent Media Commission, the Jabulani Freedom of the Airwaves conference, the then-Minister of Home Affairs Danie Schutte, IBA, the Task Group on Government Communications, COMTASK, and any relevant process, thus leading to an enabling and favourable legislative and regulatory environment. Our motivation for participating at the policymaking level was driven by the notion of democratic participation, which is rooted in the libertarian theory. Our view was that people can and must express their opinions freely, exchanging ideas without fear, and with absolute freedom, making use of technological innovations, including media platforms.

## Democratic Participation as a Critical Driver of Policies on Community Radio

In a way, the democratic participant theory is the technological version of the libertarian theory. Using technological innovations, people can and must express their views freely, exchanging ideas without fear and with freedom. Democratic Participant Media Theory is the latest in media theories, and Professor Denis McQuail is its proponent. Its location is mainly in wealthy, developed countries where the citizens have the scientific, technological, and financial means to put the latest innovations in interpersonal communication to practical and regular use.[5]

The participatory element of community radio has been enshrined in the law and its licence terms and conditions. However, it is also enabled by the radio station's founding constitution, and creative and innovative means of promoting community participation. These could be through listeners' forums, programming committees, and other innovative ways brought by digitisation, connectivity and cell phones, general community meetings, and Annual General Meetings (AGMs).

The requirement of community participation ensures that members of the community are involved in the running of the station, that is, in the following activities: ownership and control of the station, the election of leadership (board members or trustees) at a community AGM, policymaking for the station, management of the station, selection and provisioning of programming, production of programmes, and external representation of the station.

Community participation is deemed critical in the sustainability of a community radio station, as it gives meaning and effect to ownership and control by the community. It is not just a compliance issue but adds value to the business model, as the community wants to hear and listen to themselves on air; thus, it enhances listenership figures. As McQuail argues, people can and must express their views freely, exchange ideas without fear, and use technological innovations with absolute freedom. Community radio gives power to the grassroots, the right to communicate, and freedom of expression as enshrined in the Constitution Act of 1996 in South Africa.

Ideas shaped the definitions of community radio in policy, law, and regulations from communities and submissions from community organisations to the legislative and regulatory-making process. Having come from a history where some communities were not only excluded but even forbidden from having their voices heard, as activists, we were strongly motivated by the view that access to communication and information empowers citizens and assists in defending and advancing the country's democracy. The nuances in our proposals were driven by an imperative to ensure that rural communities can access all media, including radio, television, online, mobile, and print services. Contrary to the pre-1994

[5] DEMOCRATIC PARTICIPANT MEDIA THEORY – NEWSPAGES | KMA.

dispensation, where the apartheid government-controlled broadcasting promoted propaganda to defend its brutal policies and banned access to alternative information, anti-apartheid activists fought to open the airwaves and democratise the SABC. However, they demanded a platform for communities to express themselves, determine the content, and control the dialogue and narrative shaping their destiny. This is what the University of the Western Cape (UWC) initiative sought to achieve.

## UWC: The Birthplace of the First South African Community Radio Station and Media Diversity

The story of the establishment of community radio in South Africa begins in the early 1990s, when SANSCO's media subcommittee, chaired then by Satch Radebe, decided to spearhead a campaign to transform the *music jukebox*,[6] UWC Radio, into a campus community radio station. This was part of the organisation's national transformation agenda of changing campuses *From Ivory towers to People's Universities*. I was a Bachelor of Science student then engaged in youth and media activism in my spare time. As I reflect on this period, 30 years later, little did I know that these extramural activities, and not science, would be my life's course. Through these engagements, I would play some part in writing the history of media diversity in South Africa.

I was a member of the SANSCO media committee with Tango Lamani, Sifiso Ndlazi, Nobom Ncadana, Nokhanyo Madyibi, Melikhaya Vanqa, the late Duma Dube, and many others. I joined the UWC Media Society under John JJ Storom, which at the time was narrowcasting since it could not broadcast through a transmitter. Our audience were the lucky few who could listen in the student centre, and our content was primarily music and ad-libbing from DJs and presenters. Eventually, I became the society's chairperson, led its affiliation with SASPU working with among others JP Louw, and its engagement with the then Cassette Education Trust (CASET). CASET was an NGO led by the late Edric Gorfinkel, who trained us in producing radio programming. We produced cassettes on struggle activities and the Defiance Campaign led by the MDM and UDF. We played these cassettes on UWC Radio and distributed them in taxis, buses, and at rallies to provide progressive content as an alternative to the propaganda broadcast by the white-minority apartheid

[6] Jukebox is a social media player that lets you create a collaborative music room in seconds.

regime-controlled SABC. We decided to change UWC Radio to a campus-based community radio station. Our efforts were supported by the Student Representative Council, the former UWC Executive Director of Student Affairs, Saliem Mowzer, and the Rectorate led by the late Professor Jakes Gerwel and the late Archbishop Desmond Tutu.

We called the initiative Bush Radio. The word "bush" referred to UWC as a bush college. References to UWC as a bush college are rooted in its designation by the apartheid government as a second-rate teaching college, deliberately located on the outskirts of Cape Town in the bushy areas because it catered for coloured students. After several community consultations, we set up a Bush Radio Trust led by the late Advocate Essa Moosa SC, the late Vincent Kolbe, and Rashid Lombard, among others; and the Bush Radio Coordinating Committee led by Chairperson Shahied Teladia, Sue Valentine, me as treasurer, and later the chairperson. Other community members included the late Sandile Dikeni, Mervyn Swartz, and Gabriel Urgoitti. We held several Bush Radio Coordinating Committee meetings at UWC and, at times, at CASET offices in Salt River. The coordinating committee was supported by an efficient management team led by the late Edric Gorfinkel and Farah Bhoola, with the team of presenters including Luntu Mguba, Tanja Bosch, Barbara Meyer, Brenda Leonard, and Adriaan Louw.

Later, CASET "adopted" and dissolved into Bush Radio with its assets. Bush Radio was then supported by, amongst others, Frederick Ebert Stifstung, Australia (AusAid), Independent Media Diversity Trust (IMDT), and, later on, Kgaso Fund, a partnership between the South African and Danish governments, UNESCO, and many other international donors. We focused on using the broadcasting equipment to continue to produce cassettes and train people on radio broadcasting skills. Training became the strength of Bush Radio, with trainers from the African National Congress Radio Freedom like Joe Mjwara, the late Libby Lloyd, and many other international experts like Tracey Naughton. We trained people from various organisations, including the Congress of South African Trade Unions (COSATU), the South African National Civic Organisation, and other community organisations. We also facilitated workshops on how to set up a community radio station, leading to, other than the SASPU-affiliated campus radio stations (UDW,[7] Radio

[7] University of Durban Westville.

Turf,[8] UND,[9] VoW,[10] RMR,[11] Tuks,[12] etc.), several initiatives countrywide like in Soweto (SMRC[13] initiating Soweto Community Radio, now Jozi FM), Winterveldt, Soshanguve, and Radio Grahamstown, and subsequently convened a joint vision workshop at Chapman's Peak. This is where we discussed our understanding of community radio, who owns and controls a community radio, and so on. This workshop led to processes resulting in the launch of the NCRF in Soweto in December 1993.

At some stage, we got permission from UWC to apply for a broadcasting licence from the Department of Home Affairs led by then Minister Danie Schutter, who was responsible for broadcasting. Our application to broadcast was refused twice. Eventually, as part of the 1993 Defiance Campaign, we decided to pirate and broadcast "illegally" after the government refused to license us. This plan went ahead on 25 April 1993. We were visited by the South African Police, who shut Bush Radio down and confiscated its equipment. Two of our leaders, Edric Gorfinkel and Mervyn Swartz, were arrested. I was later arrested on my way back to campus, at the Salt River train station, sent to the Salt River Police Station, and later released.

Through SASPU, UWC was part of the SASPU "Freedom of the Airwaves" Campaign. With Katharine McKenzie and Caroline Green, I went to the University of Stellenbosch to submit our campaign and proposals to Professor Christo Viljoen, who headed the Viljoen Commission established by then President FW de Klerk. The commission arose following pressure on the government to transform SABC. I was also part of the national delegation to the Netherlands in August 1991 to a conference to Free the Airwaves in South Africa organised by Omroep voor Radio Freedom. UWC SANSCO, Bush Radio, Congress of South African Trade Unions (COSATU), SASPU, and so on, as part of the MDM, became central in the Campaign for Independent Broadcasting (CIB), which engaged in several protest actions, including transforming SABC from being a state broadcaster into a public broadcaster. I became the chairperson of the CIB in the Western Cape and part of leading negotiations with

[8] University of the North Radio Station.
[9] University of Durban.
[10] Voice of Wits.
[11] Rhodes Music Radio.
[12] University of Pretoria Radio Station.
[13] Soweto Media Resources Centre.

the National Party[14] government in Union Buildings for freeing the airwaves.

I was part of the CIB delegation led by Willie Currie, David Dison, and Michael Markowitz, which included, amongst others, the late Raymond Louw, that met Minister Danie Schutte at the Union Buildings in 1993. Notwithstanding our discussions with the apartheid government's Danie Schutte for leveling the playing field and creating an independent framework for licensing community broadcasters, he arbitrarily authorised some initiatives favourable to the regime, including the establishment of the right-wing Radio Pretoria and others. He did not license the first community radio, Bush Radio, and the related NCRF and SASPU-affiliated stations. The CIB and NCRF made several inputs to define community radio and add a one-year temporary broadcasting licence to ensure the broadcasting playing field was levelled during the 1994 elections. The work of the Independent Media Commission and later IBA was with input from Bush Radio, SASPU, and others, towards and after the re-establishment of the three-tier broadcasting industry in South Africa. Despite all the above very involved activities, I completed my degree in December 1993, graduated in March 1994, left UWC, resigned as Chairperson of Bush Radio, and founded the NCRF as its first CEO. That became the beginning of my work life, away from my BSc major subjects of biochemistry and physiology into the communications, broadcasting, and ICT future.

Subsequently, after my experience at UWC, we convened what we called the "Community Media 2000" conference in 1995, which led to the formation of the NCMF,[15] a body for community radio, community, print and community television activists. The Open Window Networkand Community Print Media Organisation were also established for video and television. This conference consolidated input into the Reconstruction and Development Programme, which recognised the need for media diversity and development. This was later implemented through President Thabo Mbeki's established COMTASK 1996 and the enactment of the Media Diversity and Development Agency (MDDA) and its set-up in 2003, which now leads the support for media diversity. The amendments of the IBA Act led to the issuance of temporary community radio stations,

[14] The National Party was the governing party of Apartheid South Africa from 1948.

[15] National Community Media Forum.

and the licensing of community radio grew from 1998 to today's more than 200 community radio stations licensed by ICASA.

Therefore, UWC was not only part of the establishment of the first community radio, but it has also been part of the development of broadcasting policy and regulation from the early 1990s. Through its training and workshops, it has also been involved in watering the seeds that would bud into other community stations, ensuring that as many communities as possible have access to media that allows them to get information and participate in the democratic process. Having outlined how one important player in the South African community radio landscape was established, I now turn my attention to telling the story of the work done in creating an enabling policy and regulatory environment through establishing the Media Diversity and Development Agency, the MDDA.

## Reflecting on the Beginnings of the MDDA

Besides the legislative framework created through the IBA of 1993 and now the Electronic Communications Act of 2005, which ensures that community radio exists and is licensed by the regulator, in 1996, President Thabo Mbeki established the COMTASK led by Dr Mandla Langa to review the South African communication landscape and the media industry.

With regard to the historical evidence of community contributions and tremendous international donor interest in the community radio sector, COMTASK recommended that the government should give recognition to the then Independent Media Diversity Trust (IMDT) as the interim funding administration agency for this sector. COMTASK further recommended that government match the funds provided by international donors for community media on a rand-for-rand basis, and that the state facilitate the process of setting up a statutorily recognised media development agency comprised of independently elected trustees. The agency will operate a statutorily recognised subsidy system for community and independent media in South Africa. The South African Communications Service was then transformed into the Government Communication and Information Systems (GCIS), which spearheaded and sponsored the enactment of the MDDA Act No. 14 of 2002 and the set-up of MDDA in 2003, which now leads the support for media diversity and development. The establishment of the MDDA was a considerable achievement as it added value and benefit to the sustainability of the community radio sector.

The work leading to the establishment of the MDDA was led in President Mbeki's Cabinet by the then Minister in the Presidency, Dr Essop Pahad, a passionate and visionary leader who laid a foundation for the realisation of ideals of media freedom in South Africa. Other cabinet members of the time, in the early post-1994 era, who strongly supported media diversity, include Pallo Jordan, Jay Naidoo, and the late Dr Ivy Matsepe-Cassaburi. I worked with the long-serving Minister in the Presidency, Dr Pahad, whilst I was still the Chief Executive Officer (CEO) of the NCRF, and Councillor of the IBA and ICASA. I also worked closely with the first CEO of the MDDA in 2003, the late Libby Lloyd, who set the agency systems which led to its first funding announcements in 2004 with limited resources amounting to just 10 million rand. The first chairperson of the MDDA was Khanyi Mkonza. Lloyd left the MDDA after a three-year contract in 2006, and I joined in June 2006 for eight years until June 2014. Gugu Msibi succeeded Mkonza.

The MDDA is mandated to promote media development and diversity and support and fund community and small commercial media. The MDDA took over from foreign support, the work of the Independent Media Development Trust (IMDT), and the Kgaso Fund, and thus through law supporting media diversity. In 2023 it celebrates its 20-year anniversary as an opportunity to reflect and review its value, benefit, and impact in promoting a diverse media. It also presents an opportunity for reconfiguring State-Owned Entities, looking at the possible synergies between the mandates of MDDA, the Universal Service and Access Agency, and the proposed Digital Development Fund in terms of the National Integrated ICT White Paper.

The MDDA grew its budget from 10 million rand in 2004 to more than 70 million rand in 2014 and was awarded grants worth more than 275 million rand. It prides itself on years of good governance, having been recognised and awarded several clean audit awards by the Auditor General of South Africa in terms of the Public Audit Act. This was made possible by the support and the tremendous collective leadership from the Executive Authority through Dr Essop Pahad up to the late Minister Collins Chabane and the board.

Dr Essop Pahad was an easy-to-work-with Minister, always available to support us whenever needed and ensuring we could pursue our legislated mandate without hindrance. He provided clear leadership in policy directives but understood the independence of the MDDA enshrined in the

MDDA Act. The MDDA acted only through its board without fear, favour, or prejudice.

On its tenth anniversary, the MDDA listed several deliverables and successes, including, amongst others, ten years of unqualified and clean audits, 570 supported media projects through grants, more than 247 bursaries awarded in media studies, advocating and ensuring more than 30 million rand government ad spend annually allocated to community and small commercial media, community radio commanding 25% of radio listenership, reduction of signal distribution costs, production of reference research reports on media ownership and control, production of easy to use toolkits on governance, advertising, and marketing (*Media Landscape:2014*). These summarise the value, benefit, and impact of the MDDA in its ten years of existence. In light of 20 years of its existance in 2023, its impact should be more felt in the industry. The MDDA Trends of ownership and control of media research report (2009) and Print Media Transformation Report (2013) exposed the challenges faced by the media in respect of transformation and necessitated interventions like the late Judge Langa-led Press Freedom Commission, the Press Regulation in the Media Report 2012, and the Nkwenkwe Nkomo-led Report on the Transformation of Print and Digital Media 2013. According to different published research by the Broadcasting Research Council, community radio stations such as Inkonjane FM, Radio Tygerberg, Bush Radio, and Jozi FM boast over 160,000 listeners, which is more than for some well-known commercial stations. This feat makes a compelling case for media diversity and a plurality of voices and should expedite initiatives to transform the sector.

## Conclusion

In his theory and practice of the revolution, Amilcar Cabral said, "You may have excellent ideas in your head, but if those ideas are not interwoven in the day-to-day socio-economic realities of the people, the ideas are as good as non-existent". As we celebrate 100 years of radio, we must ensure that the digital future does not lose pluralistic voices and media diversity through the essence of the three-tier broadcasting industry. A relevant and reliable communication platform is essential for the effective functioning of consultative and participatory democracy. Community radio is the voice of the voiceless and a powerful communication medium for grassroots communities.

Many training and capacity-building interventions have been carried out in the community radio sector since the 1990s. This investment has sustained the sector's growth, and many stations are celebrating more than 20 years. Several community radio activists and practitioners have since been absorbed into senior, high-ranking, and executive responsibilities by commercial or public broadcasting services, and others in government communication, in the private sector and as entrepreneurs. This is testimony to the depth of the experience generated from the industry and, therefore, further indirect investment into the economy. Some commentators regard community radio as a breeding ground for commercial and public radio and new talent. However, this must not misrepresent community radio as just a training platform; in its own right, it is a standalone sector and is recognised as such in the law.

Community media's most significant challenges range from sustainability, more so those in rural and poor areas, and staff turnover, as skilled people move to either commercial or public broadcasters. Sustainability is vital for this sector to develop and grow further. It must be looked at holistically, from community participation in ownership and control to programming, and thus the production of relevant content, to exploring options and models to fund operations like signal distribution, digitisation, and other innovative ideas to grow revenues. Progressive initiatives to complement the efforts of the national government through enabling policy and regulatory environment, interventions like the MDDA, GCIS support, and general appreciation of using community radio platforms to advertise government information and messages must be encouraged and supported.

Like in any other industry, some community radio stations are managed well, and some are not. *However, experience shows that community stations with genuine and active community participation and involvement in their ownership and control tend to be sustainable, and the community voice is heard, compared to those with little or no community participation.* Community radio listenership increases annually (9.8 million listeners as of January 2022), but the advertising, more so from the private sector, is not increasing proportionally. As a result, smaller community media organisations struggle to survive and thrive because of the lack of enough financial support from the advertising industry leading to skilled people leaving and, as such, the project struggling to retain skills. Other challenges are governance within projects, and hence our priority intervention in promoting good corporate governance.

There remains a considerable task for the industry: the development of an audience research survey, measurement, and method informed by the geographic and the massive peculiarity of the sector. In addition, it must be able to provide accurate and scientifically reliable figures of community radio listenership. Many MDDA interventions have yielded positive results for the development of the sector and the promotion of media diversity. In addition, MDDA is working on crucial research, reflecting on its 20 years, impact, needs, and media diversity index. The outcomes of this work will assist in shaping the agency's programme in the future.

One of the biggest challenges facing community media in South Africa today is that some local government communicators do not realise the importance of using it as a communication platform. For example, you would find a local meeting announcement advert in a national newspaper. There is a need for such communication to be conducted at the local level, in local media, and in the language spoken in that area. The local government must help strengthen and sustain local media. In trying to overcome these challenges, the MDDA should continue to use its partnership with the government to communicate the importance of using relevant and appropriate media. The government has also prioritised using relevant media, such as community and small commercial and public media, to communicate government messages. In addition to its contribution to the MDDA, the industry needs to consider paying some form of compensation or transfer fee when they employ talent from a community radio station, just like they do in sports.

In South Africa, communication will only be relevant when gains have been made regarding diverse ownership and control of media so that there is no skew towards the previously advantaged groups in respect of ownership in the media and lack of content diversity, as has been the case over the last 100 years. Looking ahead, the scope for improving communication in South Africa is broad considering the astuteness of the government as it established agencies like MDDA and the GCIS. We need to support their efforts and develop effective communication strategies. Further, we collectively need to create an enabling environment for the three spheres of our media, public, community, and commercial, to be sustainable and grow.

## References

Democratic Participant Media Theory – News Pages | KMA.
https://amarc.radio/about-amarc/.

https://journals.sagepub.com/doi/abs/10.1177/136787790000300205.
https://www.dailymaverick.co.za/opinionista/2015-04-30-print-media-self-regulation-lacks-effective-punishment/.
https://www.dailymaverick.co.za/opinionista/2015-08-12-sas-great-media-debate-moegsien-williams-is-mostly-right/.
https://www.dailymaverick.co.za/opinionista/2016-05-19-media-freedom-sa-needs-a-free-accountable-transformed-and-diverse-media/.
https://www.uwc.ac.za/News/Pages/UWC-Reflections-Birthing-the-First-South-African-Community-Radio-Station-and-Media-Diversity.aspx.
Gunner, L., Ligaga, D., & Moyo, D. (2011) Radio in Africa: Publics, Cultures, Communities. Wits University Press.
MDDA. (2009). Trends of Ownership and Control of Media in SA, 2009.
MDDA. (2013). Transformation of the Print and Digital Media in SA, November 2013.
MDDA Act, No. 14 of 2002.
Press Freedom Commission: Report on Press Regulation in SA, April 2012.
Report on the Transformation of Print and Digital Media, September 2013.
www.acasa.co.za.
www.aip.org.za.
www.brcsa.org.za.
www.cnisouthafrica.co.za.
www.cvet.org.za.
www.gcis.gov.za.
https://www.gcis.gov.za/sites/default/files/docs/resourcecentre/chapter_8.pdf.
www.icasa.org.za.
www.mdda.org.za.
www.nab.org.za.
www.ncrf.org.za.
www.pdmedia.org.za.
www.saha.org.za.
www.sentech.co.za.
Tettey, WJ. (2011). "Talk Radio and Politics in Ghana: Exploring Civic and (Un) Civil Discourse in the Public Sphere in Radio Radio in Africa: Publics, Cultures, Communities. (eds) Gunner, L., Ligaga, D., & Moyo, D. Wits University Press.

CHAPTER 3

# Language Revitalisation and Community Broadcasting in South Africa: A Case of Vaaltar FM

*Kealeboga Aiseng*

## Introduction

The world is currently facing a linguistic challenge in which some languages dominate others (Accurso, 2015; Nkoala, 2022). These include English, French, Spanish and Mandarin. These languages dominate in education, science, commerce, law, politics, technology, trade and economics. The global rise of these languages threatens the world's linguistic diversity; minority languages are being marginalised (Philipson, 2009). Hornberger and Coronel-Molina (2004) argue that most indigenous languages in the world are pushed aside in favour of dominant colonial languages. Hence, some indigenous language speakers experience different degrees of language loss and shift to the dominant colonial languages.

K. Aiseng (✉)
Rhodes University, Makhanda, South Africa
e-mail: K.Aiseng@ru.ac.za



S. Nkoala, G. Motsaathebe (eds.), *100 Years of Radio in South Africa, Volume 2*, https://doi.org/10.1007/978-3-031-40706-2_3

These problems call for serious intervention to save some of the world's languages facing extinction.

Popular culture, and the artefacts of popular culture, such as radio, has played a significant role in language revitalisation (Belhia et al., 2020), in particular, in Africa as newspapers and radio continue to play an essential role in uplifting the status of indigenous languages on the continent (Tshabangu & Salawu, 2021). In addition, indigenous African literature has also been paramount in revitalising African languages (Ngulube, 2012). However, the existing scholarship has not paid enough attention to how new emerging linguistic practices (Li, 2017) can also contribute to revitalising indigenous languages. As some previous scholars have stated, languages are not static; they constantly evolve (Garcia & Li, 2014).

The chapter explores how a community radio station in South Africa plays a vital role in revitalising indigenous languages. Using the case study of Vaaltar FM, a community radio station located in Taung, North West, South Africa, the chapter explores how the radio station uses translanguaging to revitalise Setswana. Translanguaging is a linguistic practice in which two or more autonomous language systems select and deploy particular features from a unitary linguistic repertoire to make meaning and negotiate particular communicative contexts (Vogel & Garcia, 2017).

Vaaltar FM is based in Taung and serves the interests of the local communities. Tswana-speaking people largely dominate Taung. This chapter demonstrates how, despite their recent addition to the South African radio landscape, community radio stations have served as sites for revitalising indigenous languages. Revitalisation here denotes a process of adopting various strategies to keep languages alive. The chapter proposes that adopting translanguaging as a form of broadcasting in community radio stations is a revitalising strategy. This is because translanguaging allows languages to come together, co-exist and enhance the community. Furthermore, this strategy eradicates radio stations relying only on some dominant or standard languages.

To understand the significance of translanguaging as a language revitalisation strategy, this chapter proposes that language, like any other living organism, has a life; it lives, dies and is reborn. Therefore, adopting translanguaging as a form of broadcasting by community radio stations is an appreciation of this analogy that (indigenous) languages are going through a constant transformation. Some elements of languages die due to many factors, such as language shift and language death. Hence, mixing indigenous languages with other languages or linguistic codes from other languages is a way of keeping them alive.

The chapter broadly discusses the notion of language revitalisation, highlighting the urgent need for this to occur with indigenous South African languages. It then considers translanguaging as a means of language revitalisation. It will then go on to look at the scholarship on South African community radio as sites of language revitalisation and provide an overview of Vaaltar FM. This station has been used as a case study to demonstrate this. This chapter attempts to fill this gap by arguing that emerging linguistic practices do not just disrupt hegemonies in multilingual spaces; they also revitalise languages (Zhou & Landa, 2019).

### *Contextualising Vaaltar FM and Its Served Community*

Vaaltar FM is a community radio station based in Greater Taung at a frequency of 93.6 Mhz. The radio station has existed since 2000 and prides itself on targeting Setswana-speaking listeners, especially young listeners. Therefore, the primary target audience of the station is people from Taung, which happens to be dominated by a Setswana-speaking group. However, the station plays across two provinces, North West and Northern Cape, in South Africa. In addition, it is also streamed anywhere on online platforms.

North West province is one of the nine provinces in South Africa. The province is multilingual, with languages such as Setswana, Sesotho, isiX hosa, isiZulu, English and Afrikaans spoken. But Setswana is the most spoken language. Radio is the most extensive means of communication in the province, and it mainly broadcasts in Setswana, English, Afrikaans and Sesotho. Some of the radio stations found in the province are the SABC South African Broadcasting Corporation (SABC)-owned Motsweding FM and Radio 2000 and community stations such as Letlhabile Community Radio, You FM and Vaaltar FM.

### *Community Radio in South Africa*

In South Africa, community radio refers to broadcasting that members of the community control to express their concerns, needs or aspirations without any external interference (Duncan & Seleoane, 1998). Fombad and Jiyane (2019) state that community radio stations account for one-third of the total number of radio listeners in South Africa and provide unmediated communication that is close to the people and important for community development. Bosch (2014) states that community media emerged in South Africa in the 1980s within the anti-apartheid discourse. Similarly, Duncan and Seleoane (1998) argue that community media, such as radio, newspapers and student media, played a significant role in the fight against apartheid in South Africa. For example, community radio stations were important for dismantling the apartheid system, empowering the majority and building a democracy that would facilitate access to information for diverse groups (Buckley, 2000).

During the 1980s, radio in South Africa was seen to have the potential to inform and mobilise communities (Bosch, 2014). Community media strongly leaned towards advocacy journalism because of their emergence within the political struggle against apartheid (Duncan & Seleoane, 1998). For this reason, Osunkuhle (2021) contends that community radio has always played a vital role as the voice of the voiceless and the marginalised communities in South Africa.

Community radio has also continued to play an essential role in post-apartheid South Africa. They exist in all nine provinces of the country and include religious stations, ethnic stations and stations that serve particular geographic areas (Fombad & Jiyane, 2019). These stations continue to play a significant social and political role in South Africa. For example, according to Fombad and Jiyane (2019), community radio remains the only accessible and readily affordable medium within the rural community and can play a significant role in the rural development of women.

Using the experiences of community radio broadcasting in the North West Province of South Africa, this chapter focuses on the role of a community radio station in revitalising Setswana. The chapter calls for awareness of how community radio stations across South Africa revitalise indigenous languages. It argues that radio in South Africa, whether it be public broadcasting, private broadcasting or community broadcasting, is vital in ensuring that the country's multilingualism and multiculturalism are realised. The chapter raises the following questions: What is the role of community radio stations in revitalising indigenous languages? Against the rise of English commercial radio stations, what is the future of indigenous community radio stations? What strategies are adopted by community radio stations to achieve multilingualism in their respective communities?

## Literature Review

### *Language Revitalisation*

Chiblow and Meighan (2021) state that colonisation and imperialism threatened and endangered indigenous languages worldwide. Resources to learn and promote indigenous languages are often severely limited (Meighan, 2021). This has led to many languages disappearing. However, while many languages worldwide continue to be eroded, new profiles of speakers are emerging because of efforts to revitalise languages (O'Rourke, 2015). Across the world, efforts by governments, political organisations, community activists and media are in place to revive languages perceived to be under threat. This process addresses some linguistic changes in society: language loss, convergence and shift.

According to Hornberger and Coronel-Molina (2004), language revitalisation implies recuperating and reconstructing features of a language or its use which are at least partially lost. However, King (2001) states that language revitalisation goes beyond efforts to expand the linguistic system of an embattled language; it also seeks to bring the languages into new domains for new uses among new types of speakers. King (2001) further reveals that language revitalisation is not necessarily attempting to bring the language back to former usage patterns; instead, it aims to bring the language forward to new users and uses. This argument sets in motion the focus of this chapter. Setswana is not an embattled language; hence, the approach towards revitalisation strategies for Setswana by Vaaltar FM is not just to save the language but to bring the language into new domains and attract new speakers to the language.

Revitalising languages typically involves expanding them into new domains or increasing their presence in existing domains (Lhawa, 2019). Some of the social institutions that have played a role in language revitalisation are television, cinema, radio, social media, education and literature. Radio has emerged as a critical domain into which languages at risk of disappearance resurface and find life again. Although many languages are under threat, radio has provided a range of opportunities for minority groups to have a presence in their languages and for these languages to continue being passed from one generation to the other.

Eisenlohr (2004) states that broadcasting in minority languages has the potential to maintain and renew these languages. McHenry (2002) and Eisenlohr (2004) argue that broadcasting in minority languages does not only encourage language maintenance and revitalisation by providing speakers with the opportunity to hear and maintain skills in these languages, but it also achieves the transformation of ideological valuations of the languages so that minority languages are also viewed as part of the contemporary world and as relevant for the future of particular social groups. In this chapter, I want to explore the critical role electronic mediation can play in revitalising indigenous languages by broadcasting using emerging linguistic practices and appreciating the linguistic transformations in these languages.

Dorian (1994) believes that in language revitalisation, the language survives; it exists, but precariously. Hence, efforts on its behalf are put into place to mobilise the remaining speakers and recruit new speakers. This argument is also relevant to the case of how Setswana is being revitalised in Vaaltar FM. The language survives, but the station still puts into place measures to attract the existing speakers of Setswana and new speakers. With the station broadcasting mainly in Setswana but still able to offer news and some content in Afrikaans and English, this is an opportunity to attract speakers of these two languages to listen to the radio station, and ultimately, they are going to be exposed to Setswana.

A primary task of language revitalisation is to revalue a subordinate language and grant it prestige by promoting the language as an emblem of ethnic or cultural identity (Friedman, 2011). However, sometimes revitalisation strategies are hindered by identity politics. Traditional native speakers may establish a social closure in revitalisation contexts that functions as an identity control mechanism, demarcating their privileged position as authentic speakers (O'Rourke, 2015). This issue can lead to frustration on the part of newcomers to the language, sometimes deterring them from using it altogether.

In the recent past, for example, the biggest radio station in the North West, Motsweding FM, hired radio presenters who were television celebrities to increase the station's listenership. Some listeners responded negatively to this strategy, citing, among other factors, that these presenters, such as Rapulane Seiphemo (a leading character in *Generations: The Legacy*), did not speak 'authentic' Setswana; they speak Setswana from the 'streets'. This negative attitude towards other dialects, codes and linguistic variables of Setswana can easily delay language revitalisation should Motsweding FM ever find itself needing to do so.

However, Vaaltar FM has never had this problem; listeners from Taung and neighbouring communities where the station reaches have never criticised the station for not using 'authentic' Setswana. What is happening at Vaaltar FM and the community's response to it is an attempt to render Setswana a universally accessible competence. Heller (2003) and O'Rourke (2015), explain that this means that instead of being valorised as a repository of authentic cultural or national identity, where language is considered 'an inalienable characteristic of group members', it becomes a resource which can potentially belong to anybody irrespective of a group membership.

Nonetheless, there are some criticisms against language revitalisation strategies. O'Rourke and Pujolar (2013) argue that language revitalisation may lead to a phenomenon of 'new' or 'neo' speakers. This denotes non-traditional speakers of a language, for example non-traditional speakers of Welsh, Irish, Gaelic and Tagalog. Similarly, Nance (2015) notes that language revitalisation may lead to new or neo speakers outnumbering native language speakers.

## *Translanguaging*

Translanguaging is a linguistic practice in which a bilingual person has an integrated linguistic system (Garcia, 2009) due to flexible bilingualism and multiple discursive practices (Karpava, Ringblom, & Zabrodskaja, 2021). In translingual practices, speakers can choose various linguistic features depending on the context in creative and critical ways (Li Wei, 2011). Translanguaging also denotes a systematic approach that unfolds using two or more languages in alternation and boosts competence and performance in all the languages (Lewis et al., 2012). Garcia (2009) further adds that translanguaging can be used spontaneously or for pedagogical purposes. The former relies on the ability of the multilingual speaker to shuttle between languages, appreciating the diverse languages that form their repertoire as an integrated system (Canagarajah, 2011). The latter refers to the support of comprehension in an educational setting (Cenoz & Gorter, 2011). The focus of this study will be on spontaneous translanguaging.

Translanguaging is vital for minority languages. Karpava, Ringblom, and Zabrodskaja (2021) contend that translanguaging can support and expand multilingualism and integrate a minority language into a broader context. Canagarajah (2011) further adds that translanguaging creates a dynamic multilingual use and enhances cultural identities. It can also generate language and metalinguistic awareness (Cenoz & Gorter, 2011).

Nevertheless, translanguaging can also be highly controversial since it can increase language change when used in more domains. This is especially true when the media, literature and education do not make conscious choices regarding language management and allow pure languages to fade out in favour of translingual practices. In this chapter, I focus on how Vaaltar FM pushes the revitalisation of Setswana by adopting translingual practices in its content. In so doing, I propose that radio stations such as Vaaltar FM can be viewed as translanguaging spaces.

Translanguaging space assumes that signs and resources are interconnected for space production (Karpava, Ringblom, & Zabrodskaja, 2021). Li Wei (2011) believes that translanguaging spaces are socially constructed contexts that allow individuals to use their linguistic resources for communication in a strategic way. With Vaaltar FM allowing presenters to adopt translanguaging as a form of communication, using translingual practices for advertisements, and enabling callers to translanguage when calling the station or even doing in-studio participation, the station transgresses and destabilises language hierarchies. This allows the station to expand the use of language in the station and invites the revitalisation of Setswana.

The practice of translanguaging is based on the idea that concepts such as language purity and verbal hygiene were constructed to support ideologies of racial, class and gender superiority (Cameron, 1995). These ideas also produced language ideologies such as linguistic dominance, language hierarchies and linguicism. However, translanguaging disrupts and interrogates these language ideologies (Flores, 2014). Furthermore, translanguaging disrupts the socially constructed language hierarchies designed to suppress minority languages (Otheguy et al., 2015). Otheguy et al., 2015) further postulate that translanguaging does this by giving attention to well-known but almost forgotten languages.

The use of translanguaging at Vaaltar FM must be juxtaposed with the prevailing perspective on language within content from the SABC. With their roots in radio stations that were established to perpetuate the notion of separateness that characterised apartheid, SABC stations that broadcast in indigenous languages espouse language purity as their approach to communication. On the other hand, some community radio stations reject the notion of language purity as it can marginalise community members who do not speak 'pure languages'. These stations can be regarded as 'political platforms' as they challenge established language ideologies. By looking at the opportunities offered by translanguaging, it is clear why radio presenters at Vaaltar FM include all linguistic resources in their engagement with listeners. Translanguaging affords these presenters strategic tools of language and agency to communicate with listeners.

However, Karpava, Ringblom, and Zabrodskaja (2021) submit that as much as translanguaging is an opportunity for minority languages, it can also threaten them. Translanguaging can save speakers from strict purist ideologies and allow people to communicate in real life in a way that does not involve a strict separation of languages (Ortega, 2019). However, translanguaging can also empower speakers of the majority languages unless some spaces are allocated to the minority languages (Cenoz & Gorter, 2019).

## Methodology

The study adopted a case study research approach. Data was collected using interviews with one of the board members of Vaaltar FM and radio station listeners. The interview with the board member was meant to gain an understanding of the radio station's history, its language choices for different programmes aired, and the presenters' use of language in various programmes. The radio station broadcasts in Setswana, English and Afrikaans. Also, presenters can incorporate other languages into their linguistic repertoires during their programmes. In addition, 16 male and female radio station listeners were interviewed for this study. This was done to assess public opinions and attitudes towards the use of language at the radio station.

The majority of the participants were approached randomly at Taung and asked to be interviewed. The board member was emailed and asked to be part of the study and their interview was done telephonically. Participant 2 is my former high school teacher that I approached because I knew that he is one of the listeners of the radio station and usually calls the station to make commentary on issues that are affecting the community.

The board member requested their identity be kept anonymous, so they are only referred to as 'the board member' in this study. The board member was asked 12 questions, and the listeners were asked five questions. All interviews were semi-structured and were conducted using both English and Setswana. The theoretical underpinnings of this study, language revitalisation and translanguaging, were used as a guide to design the interview questions. The collected data was analysed relying on the importance of radio in development (Ndhlovu, 2008) and how radio is perceived as a far-reaching medium which helps to promote the usage of indigenous languages (Mabika & Salawu, 2014). This analysis revealed that while some radio stations, especially public broadcasting radio stations in South Africa, revitalise standard indigenous languages, some community and commercial radio stations do the same by employing complexes of situated, processual and interactional communicative practices. The author could not acquire ethical clearance for the study as I had just arrived in my new institution, and I was not familiar with the ethics clearance processes.

## Discussion

In considering the station as a translanguaging space, the content, music and manner in which presenters broadcast on Vaaltar FM are all viewed as strategies for appreciating the linguistic transformations in all languages. This signifies that Setswana, like all other languages, is not stagnant; it is fluid. However, it is also a way of spreading the language even to those who do not speak Setswana. Taung is inhabited by people from different language groups and different nationalities. Some people speak Sesotho, Sepedi, Tshivenda and Xitsonga in Taung. There are also people from Zimbabwe, Mozambique, Ghana, India and Bangladesh. These people came to the area for employment, business and religion. The radio station broadcasts news in Setswana, English and Afrikaans, and plays music in these languages, and some presenters use a combination of these languages in their shows. All these are meant to revitalise Setswana, and the following subsection details how.

The collected data indicated that Vaaltar FM exhibited characteristics of a radio station involved in language revitalisation. The station board member, who also considers themselves a community activist, was asked which strategies have been implemented to revitalise Setswana. They had this to say:

> We allow our presenters to be free in how they speak Setswana; we do not control their language abilities. We have a programme that deals with the cultural and language issues of the Batswana people, and our music library consists mainly of artists who sing in Setswana. We broadcast news in Setswana, English and Afrikaans, with a chunk of it being in Setswana.

The board member further explained the motive being the establishment of Vaaltar FM:

> The station was established to promote Setswana to serve the community of Taung, empower the community, and promote Setswana as a language of pride and honour. The station, therefore, developed a set of strategies that serve these interests.

Given this information, the strategies implemented by Vaaltar FM to revitalise Setswana are part of the radio station's founding principles. The radio station exists as one of the institutions that are fighting for the protection of indigenous languages. In support of this argument, Kaschula (2006) states that radio in South Africa surpasses any other form of media when it comes to using indigenous languages. This, however, owes its legacy to the apartheid policies, which saw independent homelands creating radio stations in their indigenous languages.

Reflecting on the role of Vaaltar FM in language revitalisation, the board member continues to outline their strategies by stating the following:

> Ours is a small community radio station, but we have unique strategies to ensure that Setswana remains alive and is preserved for generations to come. We have news bulletins in English, Setswana and Afrikaans. We do the analysis of current affairs in Setswana, English and Afrikaans. We also do advertising in the medium of Setswana rather than English. But most uniquely, we also allow free-flow languages. We produce content that has a mix of languages, the same way people in Taung speak. And that alludes to the point you raised about codeswitching.

Erastus and Kebeya (2018) argue that language varieties have become extremely useful in advertising and are a ready alternative to the 'standard varieties'. The board member of Vaaltar FM has confirmed that translanguaging has become a versatile selling strategy in the station because companies are now using the practice for advertising their services. Such an embrace implies that translanguaging has become a linguistic practice of broader communication. It, therefore, shows that Setswana is growing beyond the standard varieties. This view is also put forward by Ditsele and Mann (2014), that non-standard varieties that include Sepitori, an urban variety of Tshwane, are becoming a vital communication bridge for diverse people of urban cities.

The study was also achieved with interviews with some listeners about how the radio station affects them regarding language issues. Participant 3 states:

> I love what the radio station is doing. It is giving the people of Taung the opportunity to enjoy their languages.

This opinion is also supported by participant 5:

> No one at home [Taung] speaks pure Setswana. So Vaalter is accommodating us who do not speak standard languages.

Participant 4 also supports these views and states:

> I once read a paper that talked about how young people are discouraged from reading literature such as novels, and that is because of language barriers between novels and how they speak. So, I guess what Vaalter is doing is to attract young listeners by using languages that they speak in their lives.

Participant 12, who is a teacher at a local secondary high school, states that:

> As a Setswana teacher, it is difficult to make learners learn and read novels in Setswana. The assumption is that literature in indigenous languages is boring. But at least Vaalter is giving them content in their language, but in creative ways.

What is described by the participants is also inevitable in other community radio stations that rely on youth language as a language of broadcasting. Ojwang (2017) raises the example of youth-initiated Koch FM, a model urban community radio station that primarily uses urban slang in its presentation. According to Ojwang, this station is a symbol of language revitalisation that achieves its revitalisation strategies using the youth language.

Scholars of sociolinguistics have raised these views. For example, Mensah (2016) argues that youth language in Africa reflects how young people make agentive use of language. Therefore, Vaaltar FM attracts young listeners to the station by using languages that appeal to the youth. In this way, the radio station shapes, makes and breaks the rules of engagement at will. The station uses language to conceptualise identity, enhance solidarity, foster group integration, and allow Setswana to flourish as a broadcasting language. Furthermore, the station has shown that it does not need to rely only on standard languages to achieve that. As Mensah (2016) stated, in Africa, "boundaries of youth language and mainstream language are not clear-cut: both are in steady interaction and are always renegotiated".

One of the fascinating discoveries made in this research is that Vaaltar FM is also a platform for some foreign nationals in Taung to learn Setswana. Participant 2, who is from Ghana, stated that he has been in Taung for 22 years and that Vaaltar FM helped him understand the basics of Setswana. The participant states:

> When I came to South Africa, language was a challenge, especially teaching in a rural school. I had to learn Tswana quickly. For the whole seven years, I listened to Vaalter every morning before going to work just so that I learn Tswana. I also used to hear a few Afrikaans words in the station, even though I could not understand anything.

The radio station makes Setswana appealing to those from other linguistic communities and helps them learn it quickly. These views are supported by Ugboajah (cited in Blankson, 2005), who argues that the structures and content of broadcasting systems should reflect the cultural character of the societies within which they operate. Underlying these arguments is a proposition that languages and cultures are valuable assets and that radio plays a significant role in their survival. The board member indicates that Vaaltar FM's presenters play a considerable role in the language revitalisation strategies of the station. The board member had this to say about presenters being free to express their linguistic competency:

> Unlike radio stations belonging to the SABC, we are free to use languages the way we want. And our presenters have grabbed that opportunity with both hands. We are not constrained by language policies and other political factors. We are free. We have presenters that are young and old. Some of our presenters are fluent in both Setswana and Afrikaans. And all these factors play a role in how they communicate with the listeners.

This is translanguaging, but in this case, it is not just based on how many languages an individual presenter has at their disposal or which languages are used to deliver the news. It is also about how they use all linguistic practices at their disposal to achieve communication with the listeners and keep Setswana alive and 'interesting'. Here, the radio station adopts translanguaging as a linguistic strategy to promote Setswana, protect it from possible death and revitalise it. Also, what is said by the board member alludes to strategies that connect youth and adult listeners to the radio station. There is an intergenerational approach to attracting listeners to the radio station. The radio station employs language in such a way as to accommodate both the young and the older generations. What we are witnessing in Vaaltar FM is that in situations unmonitored by governmental, educational or other authorities, people can use their entire linguistic repertoire to revitalise their languages. This radio station does not only cater for speakers who use dominant linguistic practices, but it is also welcoming to everyone. When news as a formal genre that deals with serious issues uses pure Setswana, morning and afternoon shows employ translanguaging as a communicative technique to catch the attention of young and old listeners. This technique recognises that pure languages are important socio-political categorisations but are not the only linguistic practices in multilingual societies. Radio stations and their listeners are not held at ransom to use pure languages. Radio stations continue to set free their listeners linguistically. They can also leverage their entire linguistic repertoire and the most appropriate tools to revitalise indigenous languages.

This implies that the future of Vaaltar FM in language revitalisation is bright. Participant 1 also supports linguistic creativity taking place at the radio station. They state that:

> Our identity is important as the people of Taung, and I believe Vaaltar FM is playing a central role in protecting that identity. Language is central to the identity of all people. One cannot underestimate the role of radio in protecting people's identities. (Translated by the author from Setswana to English)

It is inevitable that Vaaltar FM, as a community radio station, takes pride in promoting indigenous language use like many others in the country. Thus, the objective of Vaaltar FM is not just to serve news, entertainment and education to the communities it serves; it is also a platform through which the language, culture and identity of the people of Taung find expression. From the interview excerpts above and conversations with listeners to the station, Vaaltar FM affirms its linguistic and cultural identity in the community of Taung and its surroundings. Many community radio stations across Africa are doing the same: Fafaa FM in Uganda (Fosu & Akpojivi, 2021) and Orisun FM 89.5 (Akanbi & Aladesanmi, 2014) and MTN Radio Yoruba in Nigeria (Oyesomi & Salawu, 2019). As noted by Nwammuo and Salawu (2019), indigenous language broadcasting in Africa seems to have advanced aptitudes, as all messages transmitted in vernacular languages can deliver the message with warmth and civility. Similarly, Salawu (2006) states that using local indigenous languages in broadcasting, particularly radio, can produce anticipated outcomes better than any other media.

These arguments allude to the vital role played by community radio stations. Ojwang (2017) explains that community radio stations offer an essential intervention, especially for the African youth, because of the changes in cultures, economic stagnation and socio-political conditions. In these situations, community radio exists as a form of social adjustment. In the Vaaltar FM case, radio's role in enhancing the spread of translanguaging cannot be overemphasised. Vaaltar FM also provides an unprecedented platform for the youth to be part of the radio station. These factors flourish and spread Setswana in Taung, and everywhere Vaaltar FM reaches.

Most community radio stations invite community members to be part of their shows and ask them to take part in debates, either online or in studios. They also allow community members to call the studio and send birthday messages to their loved ones. Participant 14 states that:

> When I call Vaaltar FM, I do not have to feel like I am reading the news. I speak in any way that I want. The station does not cut my call because I speak some 'inappropriate' Setswana. (Translated by the author from Setswana to English)

This is one of the ways through which Vaaltar FM gives people with 'unconventional' linguistic practices a chance to be heard. But it is also a way of showing that Setswana is not 'universal'; many people speak different versions of it. Some speak Serolong, Setlhaping Sehurutshi, and many other dialects of Setswana. All these people must be heard. The radio station gives them that opportunity. This is one of the ways through which radio stations help people discover new cultures and protect their own cultures.

Furthermore, King (2017) states that community radio stations are often required to serve listeners' needs, especially groups under-represented or marginalised by state and commercial broadcasters. Linguistic practices that do not follow the rules of languages and that do not confine themselves to the ideology of standard languages are often excluded in South African media, especially state radio stations. Hence, community stations such as Vaaltar FM are strategically placed to fight for limited linguistic rights in the media.

During data collection, the study was also interested in whether translanguaging strategies adopted by the station did not threaten the survival of Setswana in Taung. The board member stated that:

> The station has not thrown away Setswana in its standard form. There are presenters, especially old presenters and those that have their shows in the evenings, on Sunday mornings, these presenters still use standard Setswana. We still respect the language, but we also acknowledge that there are people who do not speak standard Setswana. They must also be accommodated.

The board member further supports their strategies to revitalise Setswana and raises the following arguments:

> Languages have always changed, what we are doing might be criticised by language purists, but we are not breaking the language [Setswana]. We are adding to it. And for you as a broadcaster to know and understand what people think and want, then speak their language. (Translated by the author from Setswana to English)

It is important to note that the board member states that revitalisation strategies adopted by the radio station pose no threat to the standard form of Setswana. This is important because sometimes revitalisation strategies are efforts to eradicate standard languages. Nevertheless, radio stations involved in language revitalisation in South Africa are doing it to protect and preserve indigenous languages. Kanayama (2007) notes that community radio stations' cultural or linguistic revitalisation strategies are also crucial for building collaborative relationships among community members. As the board member states, Vaaltar FM is endorsing translanguaging to accommodate all types of listeners for the station.

The findings from this chapter are not different from its predecessors, such as Blanco and Gonzalez (2016). While focusing on the context of Europe, their research states that community media are fundamental in reinforcing cultural and linguistic plurality and diversity. As Hassen (2016) alluded to, media in indigenous languages characterises a people-orientated communication approach and the preservation of socio-cultural values, leading to cultural self-determination. In the context of this study, Vaaltar FM not only provides the community of Taung and its surroundings with news, entertainment and education, but the station also uses Setswana as a primary language of broadcasting. The station also adopts programming that uses translanguaging, allows presenters to be linguistically accessible, and encourages listeners to express themselves in languages they are comfortable with while calling into the station or participating in debates. By so doing, the station places itself as an institution that seeks to protect Setswana and revitalise it for current and new speakers.

## Conclusion

The aim of this chapter was to advance the argument on how a local community radio station in South Africa is involved in language revitalisation. My purpose was to demonstrate that in as much as radio has always been a vital player in South Africa's socio-political matters, it has also been instrumental in advancing the use of indigenous languages. The chapter acknowledges that technology can support language revitalisation efforts. Indigenous people can rely on technology, such as broadcasting, to achieve language documentation, revitalisation and promotion. Using the media to revitalise language can also encourage youth to learn their language because it is now appealing.

Mhagama (cited in Chikaipa & Gunde, 2020) states that radio stations have always focused on empowering community participation and enhancing the capacities of the local people to work together to tackle a range of social inequalities, such as discrimination and marginalisation. This chapter took its influence from indigenous radio stations as a catalyst for protecting, revitalising and preserving indigenous languages from total extinction.

This use of language in Vaaltar FM values the use of Setswana among speakers in the community of Taung, and it values such usage as a sustainable community resource that can preserve Setswana even for generations to come. However, it is also a statement against institutions such as schools that seek to treat languages as bounded entities. The use of Setswana, as is evident in Vaaltar FM, is a socio-political struggle against language policing. Therefore, Vaaltar FM promotes a positive view of multilingualism as a practical activity. It permits the speakers of Setswana in Taung and neighbouring communities to act naturally, to use and endorse languages in their everyday lives. The radio station affords the speakers of Setswana techniques on how to revive their language(s).

This chapter has contributed to understanding the role of community radio stations in language revitalisation in South Africa. However, as the study progressed, a few areas emerged that require future studies. There are three recommendations on how to take these findings forward. Firstly, this study was based on one community radio station and how it adopts translanguaging to revitalise Setswana. Further research is required on other community radio stations in South Africa to see how the findings of this study correlate with what is happening in other community radio stations. Secondly, since the findings show that languages are not rigid but rather fluid, more research is needed to show how translanguaging can be adapted to other areas of study and not just in education. Finally, in this study, it was also discovered that language revitalisation strategies and translanguaging do not pose a threat to standard languages. Instead, they merely retain existing speakers and attract new ones. Therefore, it would be worthwhile to examine a longitudinal study on the transformations in languages and look at these transformations as opportunities for languages, not threats.

## References


Accurso, K. (2015). Language Dominance/Linguistic Dominance. In S. Thompson (Ed.), *Encyclopaedia of Diversity and Social Justice*. Rowman & Littlefield.

Akanbi, T. A., & Aladesanmi, O. A. (2014). The Use of Indigenous Language in Radio Broadcasting: A Platform for Language Engineering. *Open Journal of Modern Linguistics, 2014*(4), 563–572.

Belhia, H., Moajdoubi, M., & Safwate, M. (2020). Language Revitalisation Through the Media: A Case Study of Amazigh in Morocco. *International Journal of the Sociology of Language, 2020*(266), 121–141.

Blanco, L., & Gonzalez, M. (2016). Linguistic Diversity and Communication Rights: The Role of Community in Promoting Regional or Minority Languages in Europe. *Radio, Sound and Society Journal, 1*(1), 26–41.

Blankson, I. A. (2005). Negotiating the Use of Native Languages in Emerging Pluralistic and Independent Broadcast Systems in Africa. *African Media Review, 13*(1), 1–22.

Bosch, T. (2014). Social Media and Community Radio Journalism in South Africa. *Digital Journalism, 2*(1), 29–43.

Buckley, S. (2000). Radio's New Horizons: Democracy and Widespread Communication in the Digital Age. *International Journal of Cultural Studies, 3*, 180–187.

Cameron, D. (1995). *Verbal Hygiene*. Routledge.

Canagarajah, S. (2011). Translanguaging in the Classroom: Emerging Issues for Research and Pedagogy. *Applied Linguistics Review, 2*(1), 1–28.

Cenoz, J., & Gorter, D. (2011). Focus on Multilingualism: A Study of Trilingual Writing. *Modern Language Journal, 95*(3), 356–369.

Cenoz, J., & Gorter, D. (2019). Multilingualism, Translanguaging, and Minority Languages in SLA. *The Modern Language Journal, 103*, 131–135.

Chiblow, S., & Meighan, P. (2021). Language is Land, Land is Language: The Importance of Indigenous Languages. *Human Geography, 15*(2), 1–5.

Chikaipa, V., & Gunde, A. (2020). The Role of Community Radio in Promoting Indigenous Minority Languages and Cultures in Malawi. *Journal of Radio and Audio Media, 27*(2), 1–17.

Ditsele, T., & Mann, C. C. (2014). Language Contact in African Urban Settings: The Case of Sepitori in Tshwane. *South African Journal of African Languages, 34*(2), 159–165.

Dorian, N. C. (1994). Purism vs. Compromise in Language Revitalization and Language Revival. *Language in Society, 23*(4), 479–496.

Duncan, J., & Seleoane, M. (1998). *Media and Democracy in South Africa*. Human Sciences Research Council.

Eisenlohr, P. (2004). Language Revitalisation and New Technologies: Cultures of Electronic Mediation and the Refiguring of Communities. *Annual Review of Anthropology, 33*(1), 21–45.

Erastus, F., & Kebeya, H. (2018). Functions of Urban and Youth Language in the New Media: The Case of Sheng in Kenya. In E. Hurst-Harosh & F. Erastus (Eds.), *African Youth Languages: New Media, Performing Arts and Sociolinguistics Development*. Palgrave Macmillan.

Flores, N. (2014). *Let's Not Forget that Translanguaging is a Political Act*. Retrieved February 21, 2022, from https://educationallinguist.wordpress.com/2014/07/19/lets-not-forget-that-translanguaging-is-a-political-act/

Fombad, M. C., & Jiyane, G. V. (2019). The Role of Community Radios in Information Dissemination to Rural Women in South Africa. *Journal of Librarianship and Information Science, 51*(1), 47–58.

Fosu, M., & Akpojivi, U. (2021). Is There an Opportunity for Ethnic Journalism in Indigenous Language Broadcasting in Ghana: A Study of Fafaa Radio? *Journal of Multilingual and Multilingual Development, 42*(2), 1–13.

Friedman, D. A. (2011). Language Socialisation and Language Revitalisation. In A. Duranti, E. Ochs, & B. Schieffelin (Eds.), *The Handbook of Language Socialization*. Wiley-Blackwell.

Garcia, O. (2009). *Bilingual Education in the 21st Century: A Global Perspective.* Wiley-Blackwell.

Garcia, O., & Li, W. (2014). *Translanguaging: Language, Bilingualism and Education.* Palgrave Macmillan.

Hassen, R. (2016). Language as an Index of Identity, Power, Solidarity and Sentiment in the Multicultural Community of Wollo. *Journal of Socialomics, 5*(3): 1–5.

Heller, M. (2003). Globalisation, the New Economy, and the Commodification of Language and Identity. *Journal of Sociolinguistics, 7*(4), 473–492.

Hornberger, N. H., & Coronel-Molina, S. M. (2004). Quechua Language Shift, Maintenance, and Revitalisation in Andes: The Case for Language Planning. *International Journal of the Sociology of Language, 2004*(167), 9–67.

Li, W. (2011). Moment Analysis and translanguaging space: Discursive construction of identities by multilingual Chinese youth in Britain. *Journal of Pragmatics, 43*(5), 1222–1235.

Kanayama, T. (2007). Community Ties and Revitalization: The Role of Community Radio in Japan. *Keio Communication Review, 29*(3), 5–24.

Karpava, S., Ringblom, N., & Zabrodskaja, A. (2021). Translanguaging space and translanguaging practices in multilingual Russian-speaking families. *Russian Journal of Linguistics, 25*(4), 931–957.

Kaschula, R. (2006). Indigenous Languages and the Media in South Africa. *Acta Academica Supplementum, 2*, 141–159.

King, G. (2017). History of Struggle: The Global Story of Community Broadcasting Practices, or a Brief History of Community Radio. *Westminster Papers in Communication and Culture, 12*(2), 18–36.

King, K. A. (2001). *Language Revitalisation Processes and Prospects: Quichua in the Ecuadorian Andes.* Multilingual Matters.

Lewis, G., Jones, B., & Baker, C. (2012). Translanguaging: Origins and Development from School to the Street and Beyond. *Educational Research and Evaluation: An International Journal on Theory and Practice, 18*(7), 641–654.

Lhawa, Y. (2019). Language Revitalisation, Video, and Mobile Social Media: A Case Study from the Khroskyabs Language amongst Tibetans in China. *Language Documentation and Conservation, 13*, 564–579.

Li, W. (2017). Translanguaging as a Practical Theory of Language. *Applied Linguistics, 39*(1), 9–30.

Mabika, M., & Salawu, A. (2014). A Tale of Failure: Indigenous Language Radio Broadcasting in Zimbabwe. *Mediterranean Journal of Social Sciences, 15*(20), 2391–2401.

McHenry, T. (2002). Words as Big as the Screen: Native American Languages and the Internet. *Language Learning and Technology, 6*, 102–115.

Meighan, P. J. (2021). Decolonising the Digital Landscape: The Role of Technology in Indigenous Language Revitalisation. *AlterNative, 17*(3), 397–405.

Mensah, E. (2016). The Dynamics of Youth Language in Africa: An Introduction. *Sociolinguistic Studies, 10*(1–2), 1–14.

Nance, C. (2015). 'New' Scottish Gaelic Speakers in Glasgow: A Phonetic Study of Language Revitalisation. *Language in Society, 44*, 553–579.

Ndhlovu, F. (2008). Language and African Development: Theoretical Reflections on the Place of Languages in African Studies. *Nordic Journal of African Studies, 17*(2), 137–151.

Ngulube, P. (2012). Revitalising and Preserving Endangered Indigenous Languages in South Africa Through Writing and Publishing. *South African Journal of Libraries and Information Science, 78*(1), 11–24.

Nkoala, S. (2022). Educators' Experiences of Using Multilingual Pedagogies During Emergency Remote Teaching: A Case Study of South African Universities. *International Journal of Multilingualism*, 1–14. https://doi.org/10.1080/14790718.2022.2074012

Nwammuo, A., & Salawu, A. (2019). Are Radio Programmes via Indigenous Languages the Solution? A Study of Igbo Scholars' Assessment of the Effectiveness of the British Broadcasting Corporation (BBC) in Promoting African Languages. *African Renaissance, 16*(1), 83–99.

Ortega, L. (2019). SLA and the Study of Equitable Multilingualism. *The Modern Language Journal, 103*(2019), 23–38.

O'Rourke, B. (2015). Language Revitalisation Models in Minority Language Contexts. *Anthropological Journal of European Cultures, 24*(1), 63–82.

O'Rourke, B., & Pujolar, J. (2013). From Native Speakers to 'New Speakers': Problematizing Nativeness in Language Revitalisation Contexts. *Histoire Epistemologie Langage, 35*, 47–67.

Ojwang, B. O. (2017). Reclaiming Urban Youth Identity Through Language in Kenya: The Case of Koch FM Radio. *Africa Development, 42*(2), Special Issue on Study on Oblique Identity Dynamics, 9–29.

Osunkuhle, O. (2021). Community Radio's Contribution to Political Awareness and Participation in South Africa: Lessons from Radio Grahamstown and Forte FM Radio. *Global Media Journal TR Edition, 11*(22), 1–19.

Otheguy, R., Garcia, O., & Reid, W. (2015). Clarifying Translanguaging and Deconstructing Named Languages: A Perspective from Linguistics. *Applied Linguistics Review, 6*(3), 281–307.

Oyesomi, K., & Salawu, A. (2019). Indigenous Language and MTN Radio Yoruba Advertisement: A Study of Adult Residents of Olorunda Community, Ibadan, Oyo State, Nigeria. *Gender and Behaviour, 17*(4), 14158–14168.

Philipson, R. (2009). The Tension between Linguistic Diversity and Dominant English. In T. Skutnabb-Kangas, R. Phillipson, A. Mohanty, & M. Panda (Eds.), *Social Justice through Multilingual Education*. Multilingual Matters.

Salawu, A. (2006). Indigenous Language Media: A Veritable Tool for African Language Learning. *Journal of Multicultural Discourses, 1*(1), 86–95.

Tshabangu, T., & Salawu, A. (2021). Indigenous-Language Media Research in Africa: Gains, Losses, towards a New Research Agenda. *African Journalism Studies, 43*(1), 1–17.

Vogel, S., & Garcia, O. (2017). Translanguaging. *Oxford Research Encyclopedia of Education*. https://doi.org/10.1093/acrefore/9780190264093.013.181

Zhou, S., & Landa, N. (2019). Translanguaging as Foundational Pedagogy: Disrupting Hegemonies for Academic Access in Multilingual Spaces. *Alternation, 26*(2), 290–310.

CHAPTER 4

# The Role of Community Radio in Youth Empowerment: A Case Study of Bush Radio

*John Bulani, Joseph Adebayo, and Trust Matsilele*

## Background

There are about 284 community radio stations in South Africa (The Conversation, 2022). This number shows the pervasiveness of radio in the country's young democracy, "originally designed as the 'voice of the voiceless', community radio emerged as part of the liberalisation of the airwaves in the early 1990s" (The Conversation, 2022, np). It is this central role of radio in the life of community and its impact which has been subject to intense scholarly debate for more than a century.

Writing on radio over a decade ago, Al-Hassan et al. (2011, p. 1), postulated that "radio as a communication medium has played a significant role in nations' economic, political, and cultural development". This view was further bolstered by Teer-Tomaselli and De Villiers (1998, p. 147)

J. Bulani (✉) • J. Adebayo
Cape Peninsula University of Technology, Cape Town, South Africa

T. Matsilele
Birmingham City University, Birmingham, UK
e-mail: matsilelet@cput.ac.za



S. Nkoala, G. Motsaathebe (eds.), *100 Years of Radio in South Africa, Volume 2*, https://doi.org/10.1007/978-3-031-40706-2_4

arguing that the "uniqueness (of radio) makes it one of the most appealing and universal mass media for participatory communication and action". One of the overarching reasons why radio has found resonance in communities is captured by Pavarala (2003, p. 2166) who observed that "radio is an inexpensive medium, with comparatively simple technology, and more suitable for illiterate and peasant communities and societies characterized by oral and folk traditions".

However, since the turn of the millennium with the changing political economy of media, the global financial meltdown of 2008, and also the rise of the new media technologies in the early 2000s, radio has faced increasing pressure for both funding and to survive. Where radio has managed to survive, it has adopted new media into its production practices (Matsilele et al., 2023). This iteration means radio is now more than a one-dimensional medium as consumers are now at the centre of the production practices. The changes in the global media landscape, a result of technological and economic effects, are also reflected in South Africa's media landscape. Agreeing with this view, Maphiri (2009, p. 33) intimates that "the South African media landscape has been vibrant, dynamic, and ever-changing since the dawn of democracy. It is arguably one of Africa's most prominent, complex, robust, and mature media industries". With the growing use of social media in the public square, as Bosch (2014, p. 29) rightly observes, the "use of social media has led to greater access and participation for those audiences which are already online".

While changes in production practices have seen some seismic shifts, the ownership structure has largely remained unchanged. This concern of slow or no transformation in the ownership structure is what the governing African National Congress (ANC) highlighted in its Media and the Battle of Ideas document over a decade and a half ago (2007). In the 2007 Report of the Commission on Communications and the Battle of Ideas, the ANC noted that "this offensive against our movement (by the independent media), in its content and form, is part of a global offensive against progressive values and ideas. The increasing concentration of ownership, control and content within the international media environment is reflected in the local industry". Effectively, the ANC viewed the untransformed ownership pattern as being at the centre of the anti-ANC agenda by the mainstream media.

The issue of concentration is captured by Sullivan, cited in Lloyd (2013), who notes that South Africa's media is one of the most concentrated globally, affecting the quality and the sales of its newspapers. This could be a result of the dominance in the industry by a few large media groups such as Media24 (owned by Naspers), Times Group (owned by Arena Holdings), Independent Newspapers, Caxton and the SABC. This phenomenon of most media content being controlled by a few entities is part of the colonial and apartheid legacy that curtailed the participation of people of colour in owning media. The skewed patterns in the media reflect the broader structural challenges the country faces. An analysis of ownership structures, demographics and funding models by *Mail & Guardian* (2019) shows that the boards of media houses comprise white members at 41%, 24% African, 17% coloured, 16% Indian and 2% of people from elsewhere. It is within the context of concentration and the failed project of transformation where the community media emerged to fulfil a mandate of diversifying the airwaves and aimed at creating a media landscape that reflects the needs and aspirations of all South Africans (Potye, 2018).

This chapter begins by outlining the role of community radio in social inclusion and considers the establishment of the sector in light of the 100 years of the existence of radio in this context. It then expounds on why Bush Radio is an interesting case study to consider when one examines the place of community radio in youth empowerment. A discussion on the theoretical framework employed in the study, namely Empowerment Theory, follows this before the chapter then discusses the results.

## Literature Review

Wabwire (2013, p. 39) intimates that "the historical philosophy of community radio is to use this medium as the voice of the voiceless, the mouthpiece of the oppressed people and generally as a tool for development". This view is given credence by other scholars (e.g. Fombad & Jiyane, 2019; Chemwaina, 2014; Nirmala, 2015) who regarded community radio as one of the key platforms that can fulfil the role of empowerment and the improvement of the livelihoods of community members. Community radio in South Africa grew as a response to the control of airwaves that characterised the apartheid system. Writing on this, Tacchi (2003, p. 2184) argued that "community radio *had been* introduced rapidly and recognised as playing an important role in the development of civil society in South Africa. Yet it *was* a struggling sector largely because of licensing and regulation delays, inadequate funding and skills shortages in many areas of the country". The issue of licensing remains a challenge to this day.

Community radio in South Africa is an essential component of the overall democratisation of communications that started in the 1990s in the wake of a crumbling apartheid system (Murkens, 2009, p. 1). Community radio offers a third radio model in addition to commercial and public broadcasting. These community stations serve geographic communities and communities of interest. They broadcast content that is popular and relevant to a local, specific audience but overlooked by commercial broadcasters. Despite their noble intentions and mandate, community stations face persistent challenges (NAB, 2019, p. 20), such as signal interference and technological, financial and governance support. The Parliament Portfolio Committee on Communications (Parliamentary Monitoring Group, 2017, p. 5) revealed that one of the most prominent challenges community media faces is that many government departments are not keen on advertising in their publications or radio stations. In cases where they advertise on these platforms, government departments want it done for free, an expectation they certainly do not have regarding commercial media.

Given that community media's mandate (in the case of this study, community radio) is to support the quest for community development, the challenges faced by community radio have direct ripple effects on society at large. For example, South Africa is experiencing what demographers call a "youth bulge", which occurs when young people make up a disproportionately significant percentage of the population (Goldberg, 2019) (Lujabe, 2018). According to Stats SA (2022) there are almost 10 million young people in South Africa between the ages of 15 and 24, yet only 2.5 million of them are active members of the labour force, either employed or jobless.

In 2022, the country's population rose and reached about 60.6 million people. Of them, 609,000 were aged 80 or older, while around 22.12 million belonged to the 0–19 age group. BusinessTech News (2020) reported that unemployment in the country rose to 30.1% in the first three months of 2020 compared to the fourth quarter of 2019 (Fig. 4.1).

Given rising unemployment levels, especially within the burgeoning youth population, there is an urgent need for concerted efforts to mitigate social malaise. The media, society's trusted ally, can play an essential role in radio programming to sensitise and empower. Al-Hassan et al. ( 2011) posit that community radio can play a significant role at the grassroots level for rural development. For instance, poverty, agriculture, gender inequality, education and social problems could be the focus of programming.

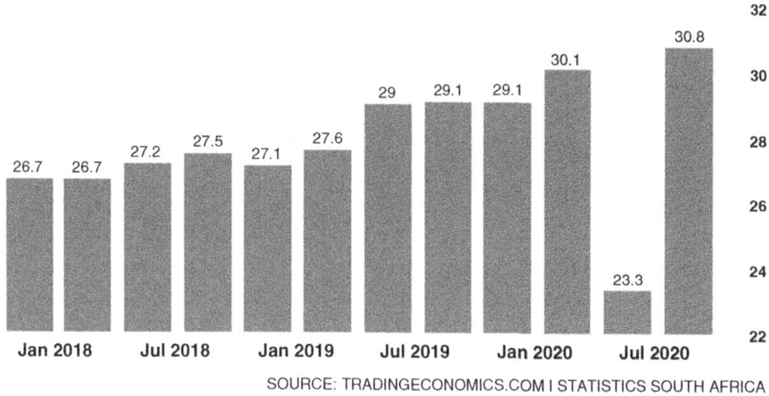


**Fig. 4.1** Two-year unemployment rates for South Africa. (Source: STATS SA, 2020)

South Africa has one of the highest inequality levels globally, with a Gini coefficient of approximately 0.67 (Arko-Achemfuor, 2016, p. 52). The World Bank reveals that youth account for 60% of Africa's joblessness. In North America, the youth unemployment rate is 25% but even more significant in Botswana, the Republic of the Congo, Senegal and South Africa (Ighobor, 2017). Italy (32%), Spain (34%) and Greece (40%) are the top three countries among the 36 OECD (Organisation for Economic Co-operation and Development) countries with high youth unemployment (Fleming, 2020).

Figure 4.2 illustrates the youth unemployment rate in South Africa between 1999 and 2020.

This graph illustrates that South African youth unemployment increased exponentially from 53.22% in 1999 to 61.04% in 2013. It indicates that the country experienced an 8% increase in five years. In summary, this graph depicts that the youth unemployment rate has been rising for many years in the last two decades and has turned out to be one of the country's foremost socioeconomic challenges. In the Western Cape, the province of interest for this chapter, youth unemployment records the highest statistic for those without matric compared to those without tertiary education and training (De Lannoy et al., 2018, p. 12). Furthermore, data shows the perpetual struggle of youth growing up in an impoverished environment, such as the Cape Flats, trying to solve their problems. In so doing, young

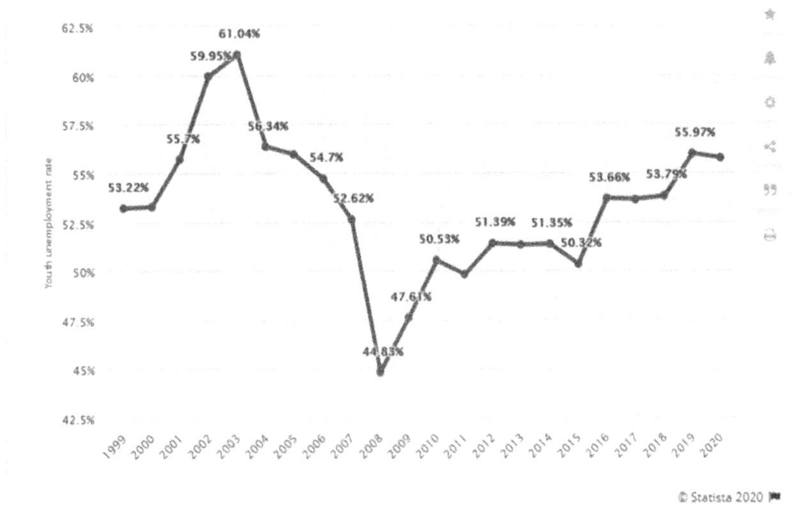


**Fig. 4.2** The youth unemployment rate in 2020. (Source: Statista, 2020)

people are asking for support and interventions that would allow them to fulfil their aspirations for themselves and their families.

With all the challenges young people encounter in South Africa, this study assesses Bush Radio's programming geared towards youth empowerment in the Cape Flats. The focus is on community radio's role in promoting youth empowerment. The media is only part of an ecosystem dedicated to ensuring that young people in any given society form part of mainstream policy decisions that affect them. Thus, it is essential to state that it is not the researchers' intention in this study to ascribe all duties for youth empowerment to community radio. The researchers understand the complexities of the youth empowerment discourse and recognise the immense role radio plays in setting public discourse agendas and initiating desirable social change.

Youth are the backbone of any society. Youth is the period of adolescence during which young people transition from childhood to adulthood, become sexually mature, and experience increasing social and economic autonomy (Sumberg et al., 2012, p. 5). There is no generally accepted definition of the term youth because of its fluidity as a category describing a period of transition from the dependence of childhood to the independence of adulthood (Adebayo, 2018, p. 146). However, age has been the best possible means of defining the group globally. For example, the United Nations defines youth thus: "For statistical purposes, the United Nations defines those between the ages of 15 and 24 as youth without prejudice to other definitions by the Member States" (Nakiyingi, 2019).

In South Africa, youth are a significant human resource for development, often acting as crucial agents for social change, economic expansion and innovation (The Presidency, 2015, p. 2). According to the National Youth Policy, as well as the National Youth Commission Act (1996), youth is defined as anybody between the ages of 14 and 35 years. Similarly, the African Union Youth Charter classifies young people between 15 and 35 as youth. However, as summarised in Table 4.1, several UN entities, instruments and regional organisations have different definitions of youth, which the United Nations secretariat recognises.

The term "empowerment" is typically viewed broadly depending on its conceptualisation (Risler et al., 2008, p. 58). According to Quinn and Davies (1999, p. 116), the concept of empowerment originated with Aristotle's (384-322 BC) conceptualisation of power as potency; "ideas and the movements of people and animals resulted from their power". Based on several definitions of empowerment from literature, it is an ongoing process to enable individuals to access and control material and informational resources. Furthermore, Eyben et al. (2008, p. 5) posit that empowerment is fundamentally anchored in power. Power allows people to have the courage to do things they never thought they were capable of doing, and comes from working alongside others to claim what is rightfully theirs. In the South African context, the government has used the

**Table 4.1** Definitions of youth according to various UN entities

| *Entity/instrument/organisation* | *Age* | *Reference* |
|---|---|---|
| UN Secretariat/UNESCO/ILO | Youth: 15–24 | UN Instruments, Statistics |
| UN-Habitat (Youth Fund) | Youth 15–32 | Agenda 21 |
| UNICEF/WHO/UNFPA | Adolescent: Youth: 15–24 | UNFPA |
| UNICEF /The Convention on the Rights of the Child | Child until 18 | UNICEF |
| The African Youth Charter | Youth: 15–35 | African Union, 2006 |

concept of empowerment for various development initiatives leading to social change, whether for youth or women's empowerment.

Against the backdrop of 100 years since the first radio service in the country in Johannesburg in 1923 (Mhlambi, 2015, p. 22), the notion that a medium that was initially introduced to audiences of colour as a tool of propaganda could eventually be reappropriated by these very audiences as a means of empowerment attests to the power of radio when it is in the hands of ordinary people to use as they see fit. The earliest manifestation of this was when indigenous language speakers used radio to express themselves in a socio-political context that had rendered them voiceless (Mhlambi, 2015). In its current form, the role of community radio to include and empower youth is yet another instance of disparate communities using it in their endeavours to claim what is rightfully theirs.

## The Role of Community Radio in the Social Inclusion of Youth

Previous studies on community radio in South Africa have focused on community radio's impact and societal role over the decades (Gondwe & Mavindidze, 2014; Bosch, 2014; Fombad & Jiyane, 2019). The sector was formally established in 1994, when the Independent Communications Authority of South Africa, ICASA, began issuing licences. When one locates its emergence in the broader timeline of the 100-year existence of radio in South Africa, one can see how of the three station types, namely public service, commercial and community, the latter are by far the most accessible to communities. Before the establishment of community media broadly, and radio in particular, marginalised groups had to jump through many hoops to have their voices heard in media. Now, despite being the most recently formed tier, it is by far the most dominant in terms of the number of radio stations and represents much wider audiences geographically, culturally, linguistically and religiously.

Marginalisation and exclusion of women and youth in society are significant issues that have existed for centuries. However, media convergence has made communication mediums more affordable and accessible to society, especially to marginalised groups such as youth, the disabled and women. This section of the study explores how community radio has been (or can be) utilised to foster young people's social inclusion through media production.

In South Africa, the Children's Radio Foundation (The Communication Initiative Network, 2011) utilises radio to create opportunities for information sharing, access to public services, dialogue, leadership, advocacy, and community building among youth. Besides offering young people training as reporters and content creators, the foundation affords them a platform to broadcast essential issues to them and their communities (giving them a voice). Some topics covered in their content are climate change, gang violence, sexual and reproductive health, and unemployment (Gupta, 2020).

Similarly, the Young Reporters Network (YRN) in Tanzania offers young people a unique opportunity to share their concerns, experiences, hopes and aspirations and effectively bring about behavioural and social change through traditional and social media content. The programme aims to increase young people's active participation in producing children's programmes and develop their capacity to advocate for children's rights (UNICEF, 2014). Greater Lebowakgomo FM is one of the proactive community radio stations in Limpopo for youth participation and involvement. The station provides young people with skills and tools to produce their weekly programmes, which serve as their platforms to share their views about young people's challenges within the station's communities.

Through these case studies, it is apparent that community radio is a preferred medium for marginalised young people to raise issues of importance to them and their communities. The content produced by these emerging broadcasters eliminates traditional stereotypes about youth and their challenges—by framing their own stories without an adult point of view. These case studies also indicate that radio is a tool that young people can use to gain experiential knowledge by producing media content about youth and their societal challenges.

Although Cape Town has more than ten on-air commercial and community radio stations, this study considered only one community radio station: Bush Radio. Bush Radio is the oldest community radio station in South Africa. Going to air on 25 April 1993, albeit unlicensed at the time, it regards itself as the *Mother of Community Radio in Africa* (Bush Radio, 2021). Its history dates to 1989, when a community service project called Cassette Education Trust began distributing cassette tapes at the height of apartheid at the time when South Africa's media industry was under the government's control (Bosch, 2008, p. 73). The station played a pivotal role in producing political content, including for the Defiance Campaign led by the Mass Democratic Movement. The rationale behind the selection of Bush Radio as a station of choice for this study is that it is the only station in Cape Town, with approximately 80% of its listeners being youth aged 15–35 (Bush Radio, 2021). Further, in light of being part of a volume that endeavours to consider particular cases concerning the century-long existence of radio in the South African context, as the oldest community radio station, it provides an interesting case study of how this tier of radio has developed over the decades.

## Theoretical Framework

This study is undergirded by the Empowerment Theory of Zimmerman (2000). This theory links an individual's well-being with the larger social and political environment (Perkins & Zimmerman, 1995, p. 569). In Gutierrez's (1995) view, the Empowerment Theory posits that people should work for the common good if they develop a critical consciousness, and claims that changes in ideas and attitudes contribute to people's participation in social change. According to Zimmerman (2000, p. 43), this theory also offers methods for measuring the construct in various situations, researching empowering behaviours, and separating empowerment from other constructs like self-esteem, self-efficacy or locus of control.

In a social work context, Rappaport (1987) refers to the word empowerment as a process by which people, organisations and communities gain mastery over issues of concern to them. The term links various activities, from individual self-assertion to collective resistance, protest and mobilisation, that challenge fundamental power relations. In Parsons (1991, p. 10), themes of conceptualising empowerment are identified as;

- a developmental process begins with individual growth and possibly culminates in more significant social change.
- a psychological state marked by heightened feelings of self-esteem, efficacy, and control.
- liberation results from a social movement that begins with education and politicisation of powerless people and later involves collective attempts on the part of the vulnerable to gain power and change those oppressive structures.

For this study, youth empowerment (from the media consumption perspective) is a process of capacitating young people ages 18–35 through information, education and entertainment to make choices and then transform those choices into desired aspirations. It also supports them in developing the knowledge and skills required to succeed. By employing this theory, the research assessed Bush Radio's role in fostering youth economic empowerment in the Cape Flats through its programming.

## Methodology

To achieve the objectives of this study, the researchers adopted a qualitative approach. In Brynard et al.'s (2014, p. 39) view, this research approach allows the researcher to interpret and describe people's actions. Furthermore, by adopting the qualitative research approach, the researchers could interact with the study respondents to gain more insight into the topic. The study employs a case study research design by undertaking an in-depth analysis of the station's role in youth empowerment programming. Semi-structured interviews with management, employees and listeners were used to gather data. The interviews aimed to identify the radio shows produced by the station and investigate how these programs support young people's empowerment.

Purposive sampling method was adopted for the study to ascertain youth perceptions of youth-focused programmes of the station. A total of 20 listeners (youth) were purposively sampled in the Cape Flats townships—Manenberg, Mitchells Plain, Khayelitsha, Elsie's River, Nyanga and Phillipi. Purposive sampling is a non-probability sampling method that occurs when the individuals selected for the sample are chosen based on the researcher's sound judgement, thereby saving time and money (Tangco, 2006). The selection of these participants was based on their gender, age and location. The participants consented to be part of the study.

## Result and Discussion

As stated earlier, we interviewed Bush Radio presenters and senior managers to determine the programmes the station produced targeting youth. As a result, the following programmes were identified as a sample for the study based on audience reach and content focus for each programme (Table 4.2).

Based on the data gathered from the semi-structured interviews, it emerged that Bush Radio has a strong representation of young people in

**Table 4.2** Bush Radio programmes targeted at youth

| *Radio programme* | *Description* | *Presenter(s)* |
|---|---|---|
| 1. *Sakh'isizwe* | Daily talk show dealing with current affairs with guests from civil society, government, political parties, entrepreneurs and community activists. The show features the local talent of artists, writers, athletes, academics and community workers. Weekly, the programme is aimed at guiding listeners in making better choices concerning daily life and information dissemination. (Monday–Friday, 12h00–14h00) | Jasmine Roberts |
| 2. *The Breakfast Rush* | The three-hour show broadcasts 50% local music and features interviews with local artists, sportspeople, entrepreneurs, fashion designers, actors and community leaders. (Monday–Friday, 07h00–10h00) | Yuzriq Meyer |
| 3. *The Morning Cruise* | Daily informative magazine programme featuring interviews on various topics, designed to help listeners get in touch with Cape Town's pulse. With *word of the day*, *Trivia questions*, *Inan fo Guide*, *Job shop*, *Tip of the day* and *Quote*, the programme is guaranteed to keep its audiences current and get them talking. The show aims to educate, enlighten, empower and inform listeners about issues concerning their daily lives without being boring (Monday–Friday, 10h00–12h00) | Mkhuseli Veto |
| 4. *Everyday People* | Drive-time show packed with insightful conversations that tackle business and finance, health, sports and even the arts. The drive-time performance ensures that the afternoon is contagious with laughs, the latest music and heartfelt conversations. After all, it is all about listeners' voices and their stories (Monday–Friday, 16h00–17h00) | Phiwe Manata and Lerato Mashile |
| 5. *The Curly Wave Show* | A weekend music show that offers a platform for young entrepreneurs, emerging and established performing artists, and DJs to promote and share their stories and content. Also, the show has a segment that focuses on the latest trends in the entertainment space (Saturday, 12h00–14h00) | Michaela Muller |

the driving seat in the production of these programmes, giving youth a voice to be heard. All presenters of these programmes are below the age of 35. The study also found that these programmes aim to inform, educate and entertain youth and the whole community served by the station. These programmes address a wide range of topics focusing on health, gender equality, entrepreneurship, unemployment, service delivery, education, arts and culture, politics, sports, and labour matters. These are some of the findings solicited from the respondents on the content of the station:

> You know, we believe that a little sugar makes the medicine go down. So, our content intersperses with social, health, political, and consciousness messages while maintaining the latest music genres played, opening an opportunity for young artists. So generally, the station's programming is geared towards a youth audience.
>
> The average age of a volunteer or producer/presenter ranges from 22 to 26. So, young people generate the bulk of the content, which goes across the board—from news to on-air content.
>
> I feel they are informative and the right programmes for our community because they promote and support small businesses.

### *Bush Radio as a Platform for Youth Economic Empowerment*

The data obtained from the semi-structured interviews have formed a basis for understanding the role played by the radio station in fostering youth economic empowerment from the station's perspective. In the context of the study, economic empowerment (from the media consumption perspective) is a process of capacitating young people through access to information, education and entertainment to make choices and then transform those choices into desired aspirations.

#### *Promotes Entrepreneurship, Skills Development, Financial Literacy and Employment*

We found that most of the programmes identified by the station have a common element of support for local entrepreneurs and they create awareness of entrepreneurship among the youth. For example, the *Everyday People* programme dedicates Mondays to local entrepreneurs, offering them a platform free of charge to promote their products and services and share their business experiences with other aspiring young entrepreneurs listening to the station. It also offers free advice on business-related matters such as company registrations, financing and taxes. In addition, every Wednesday the *Breakfast Rush* focuses on local movers and shakers in business, sports, arts, fashion and civil society. The show presents an opportunity for small local businesses to be featured to promote their products and services. Similarly, *The Morning Cruise* has a segment where youth interested in starting a business, and may need a business mentor, are connected to one in the form of experts who are invited to be part of the show. The following are some of the responses in this regard:

> Business Monday focuses mainly on businesses and entrepreneurship. We focus on businesses within the community where we spotlight them, whether you are from Khayelitsha, Woodstock, or Sea Point.
>
> Everyday People has a feature specifically around youth entrepreneurship. So, they do quite a bit, and Morning Cruise also has many segments, even the breakfast show that regularly features many young entrepreneurs.
>
> [T]he show has another aspect if you want to start your own business; we invite experts (people who are already in the business sector), and they come to the show to assist.

We further found that the station has a programme called *Sakh'isizwe,* which is centred on youth empowerment and has a segment focusing on job creation. It has a feature that offers listeners basic soft skills to enable them to find a job. These skills include crafting a CV and a cover letter and registering a business. Furthermore, the show connects unemployed youth to relevant youth organisations which create opportunities for unemployed youth. Also, data shows that *The Morning Cruise* programme has a dedicated segment that supports young people searching for employment. Finally, the *Jobshop* feature lists vacant jobs in Cape Town on-air to unemployed listeners with no internet access to online employment opportunities (advertisements) and newspaper classifieds. These views from the respondents expound on these findings further:

> Sakh'isizwe helps the youth by equipping them with employment and educational information.
>
> We have high youth unemployment. So, although certain shows do not target a youth group, things like Jobshop segments have been beneficial, and interaction is quite great with young people.
>
> The Morning Cruise—learn new words and how to upskill according to what companies are currently looking for.

As stated in South Africa's National Youth Policy, young people represent a powerful resource of the country, provided they are supported and enabled to become active members of society. In Johnson's (2007, p. 4) view, community media deals with empowerment in many ways. One of those is offering skills training programmes to build and strengthen the capacity of both its staff and volunteers. Data revealed that the station has a rich history in skills training and employment for unemployed youth interested in community broadcasting. In addition, the station offers young aspiring broadcasters, with or without qualifications in journalism or broadcasting, opportunities to gain professional on-the-job experience in presenting, producing and post-production. These interns and volunteers are not paid a salary but receive a stipend to cover their basic needs. Furthermore, the station, like many other community radio stations across the country, acts as a training ground for aspiring broadcasters who require experience in broadcasting; they then migrate to the commercial space for greener pastures.

A study by Garg and Singh (2018, p. 173) reveals low levels of financial literacy among youth across most parts of the world. In response to this challenge at the local level, the show *Everyday People* on Mondays has a dedicated segment aimed at educating youth on financial matters such as budgeting, business financial management, savings and how to invest. In addition, the show invites financial experts to engage with the listeners and capacitate them to manage their finances effectively.

One of the respondents explained it as follows:

> Know Your Story is where we get financial experts, and they breakdown financial terms because most of the time, we find that most young people don't know the terminologies in terms of what 'black tax' is, how SMMEs work, and how taxes work, and so forth. So, we normally invite a financial expert or someone within the financial space to come and breakdown some of the terms in a way people will understand.

*Promotes Local Content (Music)*

The data collected revealed that the station entrenches localism of its programming, one of the critical tenets of community media, as conceptualised by Johnson (2007, p. 3). Tucker (2013, p. 395) asserts that community media must play local music, support local artists and focus on public affairs as part of its programming principles. Localism is evident in all the youth programmes sampled for this study. As one of the regulatory requirements for community broadcasting, the station actively serves as a platform utilised by local young emerging artists to promote and monetise their creative content (music). The station fulfils this responsibility by affording local emerging and established artists a platform to engage and reach their target audiences and generate more bookings for gigs. We also found that having their music played more frequently allows these local artists to claim royalties from the South African Music Rights Organisation.

### *Youth Perceptions of Bush Radio Programming*

The data from the interviews formed a basis for understanding how youth perceive the programming produced at the station. The listeners were asked for their opinion on the role of Bush Radio as a community radio station:

> To be the voice of the community and any community-related.
>
> To act as a voice for the community, broadcast specific viewpoints of the community. In addition, share emergency information about crime hotspots on the Cape Flats.
>
> To be informative, play a role in social upliftment.
>
> Talk about what's happening locally in our marginalised communities and bring development to our people.
>
> To empower, equip and entertain the community.

These responses show that youth in the Cape Flats have some basic understanding of the normative role and function of the station under study. However, it also emerged that other respondents were not entirely knowledgeable about this.

They were then asked which programmes were their favourite and to explain this response. In this regard, some of the programmes sampled for this study were found to indeed be the preferred shows favoured by the young target audience. Below are some of the answers:

> The Breakfast Rush—because of the great music and updated news about the Cape Flats and beyond.
>
> Sakhisizwe—for it helps me to know what is happening around me and the country at large.
>
> The Morning Cruise—to learn new words and how to upskill according to what companies are currently looking for.

The youth surveyed were asked for their reflections concerning their awareness of programmes on Bush Radio that specifically deal with the issue of youth empowerment. The table provides an overview of their responses in this regard.

| *Study areas* | *Yes* | *No* | *Not sure* |
|---|---|---|---|
| 1. Khayelitsha | 2 | 0 | 2 |
| 2. Manenberg | 1 | 0 | 1 |
| 3. Nyanga | 1 | 1 | 1 |
| 4. Mitchells Plain | 1 | 2 | 2 |
| 5. Phillipi | 2 | 1 | 1 |
| 6. Elsie's River | 1 | 1 | 0 |
| Total | 8 | 5 | 7 |

We found that 40% of the respondents are aware of the programmes designed by the station focusing on youth empowerment, whereas 25% are not aware, and the other 35% are unsure if these programmes exist. The findings indicate that the respondents have good impressions of various programmes produced, and each respondent has a different perspective on each programme. Some of the feedback received in response to this question posed is as follows:

> It helps the youth by equipping them with employment and educational information.
>
> I'm just enjoying the music.
>
> Listeners share what they think, and the community get a sense of what's happening concerning local developments.
>
> It needs more local music and adds more content.
>
> They benefit the community.

When asked if they would say that Bush Radio is indeed fostering youth economic empowerment, the responses revealed that most respondents affirmed they believe that the station plays a role in youth economic empowerment through its programming:

> Yes, because if you listen to the programme, you end up getting that job you were probably looking for.
>
> Yes, programs are educational and entertaining.
>
> Yes, it is trying, but it must try harder.
>
> Sort of, as it helps the youth know which industries demand which skills.
>
> Yes, because its promotions [are directed] towards young businesses.

An important way in which community radio stations have empowered listeners over the decades is by providing them with platforms to air their views and have a say in programming and content decisions. As such, the youth were asked about the extent to which they believe they have avenues to make suggestions to the station. The table below provides an overview of their responses.

| *Study areas* | *Yes* | *No* | *Not sure* |
|---|---|---|---|
| 1. Khayelitsha | 1 | 2 | 1 |
| 2. Manenberg | 0 | 0 | 2 |
| 3. Nyanga | 1 | 2 | 0 |
| 4. Mitchells Plain | 1 | 2 | 2 |
| 5. Phillipi | 1 | 2 | 1 |
| 6. Elsie's river | 0 | 1 | 1 |
| Total | **4 (20%)** | **9 (45%)** | **7 (35%)** |

Data shows that most (45%) of the respondents are unaware of channels at the station to add or forward their suggestions and inputs regarding programming. A minimum number (20%) of the respondents are aware of the avenues created by the station for advice and views from the listeners. Another 35% are not sure of these channels. According to Nirmala (2015, p. 42), radio stations such as Bush Radio must provide a voice for marginalised groups in society, such as women, the disabled and youth, to express their views on issues facing the community served by the station. To ensure relevance to its listeners and the community it serves, Alfandika (2019) recommends that a community station forms a relationship anchored in two-way communication where access and participation of the community in the station are paramount.

Finally, the respondents shared their insights when asked for suggestions on how the station can foster youth economic empowerment through its programmes:

> Running youth workshops for entrepreneurs and work together with schools on upliftment programmes in poor communities.
>
> Bringing in youth that is successful in their respective fields to share their journey in the world of work, to inspire the fellow South African sitting at home without hope and unemployed.
>
> Be engaged with community forums, team up with schools, police, and NGOs to become proactive in youth activities.
>
> Share business details and give them [youth] the opportunities to express their business ideas.

By giving youth advice on their lives. Also, help them pursue their goals as well as their dreams.

## Conclusion

The study found that the Bush Radio has several programmes that cater to the youth in the delineated areas. Bush Radio has a strong representation of young people in the driving seat in producing these programmes, giving youth a voice that is heard. Aligned with the ICASA mandate for community broadcasting, the study also found that the station's programming aims to inform, educate and entertain the youth and the whole community served by the station. These programmes address a wide range of topics focusing on health, gender equality, entrepreneurship, unemployment, service delivery, education, arts and culture, politics, sports and labour matters.

In response to the main research question on the role of the community in youth economic empowerment, the study found that the station offers a wide range of mechanisms to empower its listeners (especially youth). Firstly, it has been found that most of the programmes sampled for this study have a common element of support for local young entrepreneurs and create awareness of entrepreneurship amongst the youth. Through these identified programmes, young emerging entrepreneurs can be offered a platform to promote their products and services at no cost, compared to the enormous costs they would incur with a commercial station. Secondly, the station provides young aspiring broadcasters with opportunities to gain professional on-the-job experience in presenting, producing and post-production. These interns and volunteers are not paid a salary but receive a stipend to cover their basic needs.

Furthermore, the data gathered indicates that the station also caters to young people sitting at home with or without formal qualifications in need of a job. Through programmes such as *Sakh'isizwe* and *The Morning Cruise*, the station provides listeners with relevant lists of vacant positions in Cape Town. Also, these include sharing information on work preparedness (how to draft a CV, cover letter, etc.), financial literacy, bursaries, internships and leadership.

The study sought to understand youth's perceptions regarding the programming the station designed. A questionnaire was distributed and administered. Firstly, most respondents were aware (informed) of the normative mandate of community stations such as Bush Radio to the geographical communities it serves. However, it was also discovered that many of those surveyed were not listening or even aware that the station existed in their area. Data shows that the programmes sampled were favourable and relevant to the youth. Also, the study has found that 40% of the respondents sampled were aware of the programmes designed by the station focusing on youth empowerment, whereas 25% of the respondents are not aware, and the other 35% are not quite sure if these programmes exist. Data also showed that most (45%) respondents were unaware of channels at the station to add or forward their suggestions and inputs regarding programming. A minimum number (20%) of the respondents are aware of the avenues created by the station for advice and views from the listeners. Another 35% of listeners are not sure of these channels.

As radio in South Africa marks 100 years this year, traditional listenership has been growing for decades. In addition, the broadcasting environment has experienced more people tuning in online and engaging with radio stations in various ways, such as through social media platforms. This study has discovered that the same approach has been adopted by Bush Radio in engaging with its targeted audience (youth). Furthermore, these findings illustrate that radio as a medium over the past century has remained a viable tool for educating listeners and informing them of important issues that can improve their knowledge. Also, it has been discovered that despite all the changes and improvements in the South African broadcasting environment since the dawn of democracy, community radio is still a reliable source of information for many South Africans in rural and disadvantaged areas such as the Cape Flats.

## References

Adebayo, J. (2018). Gerontocracy in African Politics: Youth and the Quest for Political Participation. *Journal of African Elections, 17*(1), 140–161.

Alfandika, L. (2019). Democratising Communication: Media Activism and Media Reform in Zimbabwe. Unpublished PhD thesis submitted to the University of the Witwatersrand, Johannesburg.

Al-hassan, S., Andani, A., & Abdul-Malik, A. (2011). The Role of Community Radio in Livelihood Improvement: The Case of Simli Radio, Field Actions Science Reports. *The Journal of Field Actions, 5.*

ANC. (2007). Report of the Commission on Communications and the Battle of Ideas. https://www.anc1912.org.za/3rd-national-policy-conference-report-of-the-commission-on-communications-and-the-battle-of-ideas/

Arko-Achemfuor, A. (2016). Addressing the challenges of food security and youth unemployment in South Africa through land reform policies. *Environmental Economics, 7*(3), 52–59.

Bosch, T. (2008). Kwaito on Community Radio: The Case of Bush Radio in Cape Town, South Africa. *The World of Music*, 75–89.

Bosch, T. (2014). Social Media and Community Radio Journalism in South Africa. *Digital Journalism, 2*(1), 29–43.

Brynard, D., Hanekom, S. X., & Brynard, P. (2014). *Introduction to Research.* Van Schaik.

Bush Radio. (2021). Bush Radio Station History. Retrieved May 23, 2021, from https://bushradio.wordpress.com/about

BusinessTech News. (2020). South Africa's unemployment rate climbs to 30.1%. Retrieved April 18, 2022, from https://businesstech.co.za/news/government/409897/south-africas-unemployment-rate-climbs-to-30-1/

Chemwaina, S. (2014). The Role of Community Radio Stations in Enhancing Rural Development; A Case Study of Kass FM. *Journal of Scientific Research and Essay, 3*, 10–15.

De Lannoy, A., Fortuin, A., Mpofu-Mketwa, T., Mudiriza, G., Ngcowa, S., Storme, E., & Smith, C. (2018). *Unpacking the Lived Realities of Western Cape Youth.* Exploring the well-being of young people residing in five of the most deprived areas in the Western Cape Province, research summary, Cape Town: Department of the Premier: Western Cape Government and Southern Africa Labour and Development Research Unit, University of Cape Town.

Eyben, R., Kabeer, N., & Cornwall, A. (2008). Conceptualising Empowerment and the Implications for Pro-poor Growth. *DAC Poverty Network by the Institute of Development Studies,* Brighton.

Fleming, S. (2020). Three big challenges for young people today. Retrieved June 20, 2020, from https://www.weforum.org/agenda/2020/01/young-people-youth-challenges-life-expectancy-.employment-homes

Fombad, M. C., & Jiyane, G. V. (2019). The role of community radios in information dissemination to rural women in South Africa. *Journal of Librarianship and Information Science, 51*(1), 47–58.

Garg, N., & Singh, S. (2018). Financial Literacy Among Youth. *International Journal of Social Economics, 1*(2), 1–18.

Goldberg, M. (2019). How to Prevent a Demographic "Youth Bulge" From Causing Widespread Unemployment. Retrieved April 24, 2023, from https://gdc.unicef.org/resource/how-prevent-demographic-youth-bulge-causing-widespread-unemployment

Gondwe, T., & Mavindidze, Z. (2014). An exploration into the operations and sustainability of community radio stations in South Africa. *Occasional Paper Series, 1*(1), 1–5.

Gupta, N. (2020). How podcasting is developing in Africa. Retrieved March 20, 2020, from https://themediaonline.co.za/2020/10/how-podcasting-is-developing-in-africa

Gutierrez, L. M. (1995). Understanding the Empowerment Process: Does Consciousness Make a Difference? *Social Work Research, 19*(4), 229–237.

Ighobor, K. (2017). Africa's Jobless Youth Cast a Shadow Over Economic Growth. Retrieved May 30, 2021, from https://www.un.org/africarenewal/magazine/special-edition-youth-2017/africas-jobless-youth-cast-shadow-over-economic-growth

Johnson, F. (2007). What Is Going on in Community Media? *Benton Foundation.* https://www.benton.org/headlines/what%E2%80%99s-going-community-media

Lloyd, L. (2013). South Africa's Media 20 Years After Apartheid. *Centre for International Media Assistance.*

Lujabe, N. (2018). At what age does the concept of 'youth' end? It depends on who you ask. Retrieved April 17, 2021, from https://www.news24.com/citypress/voices/at-what-age-does-the-concept-of-youth-end-it-depends-on-who-you-ask-20180615

Mail & Guardian. (2019). Who Runs SA's Media Is a Black-and-White Issue. https://mg.co.za/article/2019-01-04-00-who-runs-sas-media-is-a-black-and-white-issue/

Maphiri, N. (2009). *A Case for Community Media*. Shereno Printers.

Matsilele, T., Makwambeni, B., Nkoala, S., & Bulani, G. J. (2023). Youth Audiences and Social Media Integration in Community Radio Stations in South Africa: A Case Study of Zibonele FM and Bush Radio. In C. Radio (Ed.), *Youth and Urbanity in Africa: Emerging Trends and Perspectives* (pp. 123–153). Springer International Publishing.

Mhlambi, T. N. (2015). *Early radio broadcasting in South Africa: culture, modernity & technology.*

Murkens, R. (2009). The 'special interest' and 'geographic' Models of Community Radio: A Study of the Effectiveness of the Two Models in Meeting the Community's Needs. Doctoral dissertation, Department of Journalism and Media Studies, the University of Witwatersrand.

Nakiyingi, M. A. (2019). *An analysis of youth oriented built environment design in Uganda: The case of youth centres in Kampala*. Doctoral dissertation, Makerere University.

National Association of Broadcasters. (2019). State of the Broadcasting Industry Report 2015–2018: 25 Years of Freedom in Broadcasting. 2nd ed.

Nirmala, Y. (2015). The Role of Community Radio in Empowering Women in India. *Media Asia, 42*(1), 41–46.

Parliamentary Monitoring Group. (2017). Report of the Portfolio Committee on Communications on its oversight visit to KwaZulu-Natal province on 27 to 29

June 2017, dated 17 October 2017. Retrieved April 2022, from https://static.pmg.org.za/171017pccommreport.pdf

Parsons, R. J. (1991). Empowerment: Purpose and Practice Principle in Social Work. *Social Work with Groups, 14*(2), 7–21.

Pavarala, V. (2003). Building Solidarities: A Case of Community Radio in Jharkhand. *Economic and Political Weekly*, 2188–2197.

Perkins, D. D., & Zimmerman, M. A. (1995). Empowerment Theory, Research, and Application. *American Journal of Community Psychology, 23*(5), 569–579.

Potye, Z. (2018). Taking community media into the future. Retrieved May 19, 2022, from https://www.sowetanlive.co.za/opinion/columnists/2018-08-13-taking-community-media-into-the-future/

Quinn, J., & Davies, P. (1999). *Ethics and Empowerment*. Purdue University Press.

Rappaport, J. (1987). Terms of Empowerment/Exemplars of Prevention: Toward a Theory for Community Psychology. *American Journal of Community Psychology, 15*(2), 121–148.

Risler, E., Holosko, M., & Hatcher, S. (2008). A Conceptual Model for Empowering Youth Through Volunteerism. *International Journal of Volunteer Administration, 25*(1).

Statistics South Africa. (2020). Vulnerability of Youth in the South African Labour Market. Retrieved April 19, 2021, from http://www.statssa.gov.za/?p=13379

Statistics South Africa. (2022). South Africa's youth continues to bear the burden of unemployment. Retrieved March 20, 2022, from https://www.statssa.gov.za/?p=15407#:~:text=South%20Africa%20has%20over%2010,labour%20force%20(i.e.%20inactive)

Sumberg, J., Anyidoho, N. A., Leavy, J., Te Lintelo, D., & Wellard, K. (2012). Introduction: The Young People and Agriculture 'problem' in Africa.

Tacchi, J. (2003). Promise of Citizens' Media: Lessons from Community Radio in Australia and South Africa. *Economic and Political Weekly*, 2183–2187.

Tangco, M. D. C. (2006) Purposive Sampling as a Tool for Informant Selection. *Ethnobotany Research and Applications, 5*.

Teer-Tomaselli, R., & De Villiers, C. (1998). Radio: Theatre of the Mind. *The South African Handbook on Mass Communication.*

The Communication Initiative Network. (2011). Children's Radio Foundation. Retrieved February 20, 2021, from https://www.comminit.com/unicef/content/childrens-radio-foundation-0

The Conversation. (2022). Radio is thriving in South Africa: 80% are tuning in. Retrieved March 20, 2021, from https://theconversation.com/radio-is-thriving-in-south-africa-80-are-tuning-in-176846

The Presidency of the Republic of South Africa. (2015). National Youth Policy 2015–2020. Retrieved April 15, 2021, from https://www.gov.za/sites/default/files/gcis_document/201610/nationalyouthpolicy.pdf

Tucker, E. (2013). Community Radio in Political Theory and Development Practice. *Journal of Development and Communication Studies, 2*(3), 392–420.
United Nations International Children's Emergency Fund. (2014). Young Reporters Network in Tanzania. Retrieved April 7, 2021, from https://www.youtube.com/watch?v=5t8wX2zz0nQ
Wabwire, J. (2013). The Role of Community Radio in Development of the Rural Poor. *New Media and Mass Communication, 10*, 40–47.
Zimmerman, M. A. (2000). Empowerment Theory. In *Handbook of Community Psychology* (pp. 43–63). Springer.

CHAPTER 5

# Radio Waves of Inclusion: A Practitioner's Perspective on the Catholic Church's Use of Radio Broadcasting for Evangelisation in South Africa

*Freeborn Isaac Kibombwe*

## Introduction: The Catholic Church and Its Reflection on Media in General

Although religious radio stations are a fairly new addition to the South African radio landscape, having emerged as part of the mushrooming of community media in the 1990s in this context, the broadcasting of religious content dates far back. Globally, the Christian community radio stations, popularly known as media ministries, are said to have emerged in the United States in the 1930s, when "Reverend Father Coughlin started his militantly anti-communist radio broadcasts" (Osunkunle & Wozniak, 2015, p. 83). In the South African context, one of the first community

F. I. Kibombwe (✉)
Cavendish University of Zambia, Lusaka, Zambia
e-mail: fkibombwe@sjti.ac.za



S. Nkoala, G. Motsaathebe (eds.), *100 Years of Radio in South Africa, Volume 2*, https://doi.org/10.1007/978-3-031-40706-2_5

radio stations to start broadcasting was Kingfisher FM, a Christian radio station based in Port Elizabeth, in 1995. In discussing the importance of the Catholic Church's contribution to the media in South Africa in the context of the 100th year celebrations, this chapter highlights why the Church attaches such importance to these means of communication. Various documents have been written by individual popes as well as by the Pontifical Council for Social Communication (PCSC).[1] In general, the importance of Catholic media cannot be over-emphasised. Through various documents that will be cited, this study highlights why this is the case and elaborates on the purpose of the media for this religious community.

It begins by outlining the Catholic Church's understanding of media in general, focusing on the key documents or corpus of Catholic media. It then examines the Pope's document, *Miranda Prorsus*, which broadly focuses on cinema, radio and television, before homing in on what the text says about radio. From the bird's eye-view of media and Catholicism, it considers Catholic radio in South Africa through a consideration of Radio Veritas and its current contribution to South African Society as a particular faith media.

Besides *Miranda Prorsus*, four key texts have informed Catholic radio in South Africa: *Vigilanti Cura, Inter Mirifica, Communio et Progressio* and *Aetatis Novae*. Below is an elucidation of the documents and their importance in the promotion of Catholic media in general.

## Vigilanti Cura *("Be Vigilant")*[2]

This document[3] was written by Pope Pius XI in 1936 as a response to what was happening in the United States regarding development of the film industry. The clergy and ordinary people felt that the movies were becoming too expressive with language and even nudity for actors and actors, and hence the reaction. The Pope reminded the film industry then:

[1] This Council has since been replaced by the Dicastery in the Office of the Pope at the Vatican.

[2] Note that all documents by the Pope are first written in Latin and then translated into various languages. The title of any document written by the Pope comes from the first two words in the first sentence of the document. In this case the key words are *Vigilanti Cura* (Be Vigilant). Usually in terms of citation all documents have a number or each paragraph has a number; hence from now on the documents will be quoted by the abbreviation of the document title, for example Vigilanti Cura (VC #...).

[3] Most documents written by the Pope are technically known as encyclicals or letters.

> But the matter is of so great weight, especially if we consider the condition of society in our times, that it seems to Us opportune to set forth the matter again more plainly by these Letters, imparting the precepts which the times require not only of you, Venerable Brethren, but of all the Bishops of the Catholic world. It is indeed absolutely necessary and of great urgency to provide that whatever of God's gifts the progress of the age may have added either to human learning or to technical and scientific skill, may so serve divine glory, the salvation of souls and the furthering of the reign of Jesus Christ, that all of us, in the words in which the Church bids us pray, 'may so pass through temporal goods that we should not lose the eternal'. (VC 6)

The Pope highlights that with the development of the film industry, challenges also abound in the content of these films. This was some decades ago, but the challenge of what some describe as morally corrupt films continues to be produced by various film producers and directors without consideration of the ethical implications. It is in this light that many Catholics in the United States who influenced the writing of this document with the support of priests and bishops asked producers to sign a "praiseworthy" agreement; this, however, was being abrogated as the document states that:

> Nonetheless, though they had entered into this praiseworthy agreement, the producers and the operators seemed not to be able or to be unwilling to fulfil their voluntary obligations. Since therefore this agreement was shown to be of so little value, and the cinema continued to exhibit scenes of vice and crime, the way to decent entertainment by the means of the cinema seemed almost to be closed. (VC 9)

The Pope further indicated that:

> The business of the production of cinema films which conform to good morals would be at once concluded if it were possible to produce pictures in accordance with Christian principles. Wherefore we shall always continue to praise those who have in the past, or who will in the future devote themselves to this art, with the intention of producing films designed to promote healthy education and Christian principles. (VC 31)

This clarifies that promoting good education and Christian principles and morals is always the goal of Catholic media and this must always be borne in mind. However, this does not mean judging, but rather giving guidance to society at large.

### Inter Mirifica *(Decree on the Means of Social Communication)*

This document was written in 1963, not by an individual Pope, but by many bishops of the Catholic world and brothers from other faiths who were invited to attend in an ecumenical spirit the Second Vatican Council in Rome. In this document, several aspects are emphasised that talk about the essence of the Catholic media in general but apply to particular media, in this case radio, that are being discussed. The following are the assumptions for Catholic media as outlined in this document:

> Man's genius has with God's help produced marvellous technical inventions from creation, especially in our times. The Church, our mother, is particularly interested in those which directly touch man's spirit and which have opened up new avenues of easy communication of all kinds of news, of ideas and orientations. Chief among them are those means of communication which of their nature can reach and influence not merely single individuals but the very masses and even the whole of human society. These are the press, the cinema, radio, television and others of a like nature. These can rightly be called 'the means of social communication'. (IM 1)

The Church knew that even today these means of communication as developed have an advantage for the whole human community, but at the same time, they also pose some form of danger in that we use if not abuse them. It is in this light that the document further indicates:

> The Church, our mother, knows that if these media are properly used they can be of considerable benefit to mankind. They contribute greatly to the enlargement and enrichment of men's minds and to the propagation and consolidation of the kingdom of God. But the Church also knows that man can use them in ways that are contrary to the Creator's design and damaging to himself. Indeed, she grieves with a mother's sorrow at the harm all too often inflicted on society by their misuse. (IM 2)

The Catholic Church's position on the media indicates that the instruments of social communication are very good; however, they must answer to the "moral order" question. This is the essence of having media in all forms besides being tools of evangelisation; the Church ought to check how they promote the moral dimension of our society. This is in comparison to some contemporary media that can air whatever they want without any accountability; for those involved in Catholic media, the question of morality must be implied in their establishments and made a priority.

## Communio et Progressio *(on Unity and Advancement)*

This document was written in 1971 following the publication of *Inter Mirifica* as requested by the Pope, the bishops and all those who attended this influential Council (Second Vatican Council) where several themes were discussed, media being among them. The document fell under publication by the Pontifical Council for Social Communication (PCSC) at the Vatican. The opening lines remind Church media practitioners of the appreciation of how these means of communication have created unity through the advancements of technology. Hence, the document states:

> The Unity and Advancement of men living in society: these are the chief aims of social communication and of all the means it uses. These means include the press, the cinema, radio and television. The constant improvement in the media puts them at the disposal of more and more people who in their daily lives make increasing use of them. More than ever before, the way men live and think is profoundly affected by the means of communication. (CP 1)

Further counsel is given that:

> This Instruction, it is hoped, will be well received by all those who are professionally involved in the field of communications and, indeed, by all who, from goodwill, seek the progress of mankind. So, as a result of exchanges of views and cooperation with such men, the vast potential that lies in the means of social communication will be made good and this for the advancement of all. (CP 5)

The emphasis in this regard is to ensure that media is there to advance life as human beings but at the same time to always resonate with the good and advancement of human life, not the opposite. This document reminds us that technological advances are there to bring about good and unity.

## Aetatis Novae *(at the Dawn of the New Era) Commemorating the Twentieth Anniversary of* Communio et Progressio

Like the previous document, *Communio et Progressio* was written by the Pontifical Council for Social Communication (PCSC) in 1992. This was to commemorate the 20th anniversary of *Communio et Progressio*, and at the same time, to highlight several issues related to the Church and pastoral communication, and how we live out these complicated instruments of communication in our daily life with moral obligations. Therefore, the opening line, true to the document title, states:

> That at the dawn of a new era, a vast expansion of human communications is profoundly influencing culture everywhere. Revolutionary technological changes are only part of what is happening. Nowhere today are people untouched by the impact of media upon religious and moral attitudes, political and social systems, and education. (AN 1)

The reminder that in this light the technological revolution comes with its baggage is applicable in that:

> As more than just a technological revolution, today's revolution in social communications involves a fundamental reshaping of the elements by which people comprehend the world about them, and verify and express what they comprehend. The constant availability of images and ideas, and their rapid transmission even from continent to continent, have profound consequences, both positive and negative, for the psychological, moral and social development of persons, the structure and functioning of societies, intercultural communications, and the perception and transmission of values, world views, ideologies, and religious beliefs. (AN 4)

The document further reminds the reader that all forms of media must respect culture and that:

> [E]ven as the Church takes a positive, sympathetic approach to media, seeking to enter into the culture created by modern communications in order to evangelize effectively, it is necessary at the very same time that the Church offer a critical evaluation of mass media and their impact upon culture. (AN 12)

## The Catholic Church and Radio—A Focus on Pope Pius XII's Encyclical *Miranda Prorsus* ("Wonderful Inventions")

The publication of the Encyclical (letter) *Miranda Prorsus* (MP) in 1957 by Pope Pius XII was a turning point in the Catholic Church. It presents a clear statement on the role of the media in society and what was expected of the Catholic Church membership and the world at large. The letter focused on three key media platforms that had taken the world by storm early on: motion picture (cinema), radio and television. This section will focus on how the Pope presented the role of radio and how the Church can embrace these modern developments for evangelisation and promoting human dignity. In the general introduction Pope Pius XII writes:

> The truly wonderful invention of science, the cause of so much pride to men [sic] of our day, are admittedly the fruits of man's [sic] genius and labours. But they are also God's gifts to man [sic]; the gifts of our Creator, who is Himself the Author of all that is well done. (MP 1)

The Pope calls these "wonderful inventions" because he has been remembered as the pope of the modern media, having written extensively about science and developing technologies. This indicates that anyone involved with the Church must make these inventions a priority through his urging. Sometimes media advancement is not prioritised in our so-called Pastoral Planning for the Church due to a lack of interest or proper preparation in basic communication studies. However, one can be consoled by the fact that other documents would be written after *Miranda Prorsus*. In looking at the overall picture of this document that talks about cinema, radio and television, the Pope succinctly indicates the purpose of these media platforms and the role they play in the Church and society by stating:

> Some of these inventions enhance and redouble man's physical power Others serve to better the conditions of his life. Others again have their most powerful effect upon his mind. These are designed either directly or through the medium of the synthetic pictures and sounds which they produce, to reach the mass of the people and to communicate to them messages and ideas and precepts of every description and with the utmost facility. They provide as it were, a mental food with which the people may be nourished during their hours of rest and relaxation. In this final category the three inventions which have made the most outstanding progress in our lifetime are the cinema, sound broadcasting, and television. (MP 2)

## Pope Pius XII and Sound Broadcasting (Radio)

In addressing this means of communication, the Pope indicates that this form of media is essential in passing on information due to its popularity. From the outset, the Pope reminds us of the address he gave to a certain radio society in the following words: "the position (of sound broadcasting) is as follows: it is, as it were, dissociated and immune from those limitations of time and place which handicap and retard all other human means of communication. By its means messages are transmitted in airborne flight faster than the speed of sound, swift as the passage of light, over leaping all boundaries in a moment of time" (MP 82).[4]

Radio was seen and is still seen as a blessing that has helped change the world and how information moves from one point to another. This medium is affirmed by the Pope declaring, "the radio brought by recent discoveries to near perfection, is of exceptional benefit to the technical arts in general" (MP 83). He further adds that, "to be able to sit at home and yet to hear what other men are saying, to know what is going on at a distance, and to share in the advantages of social life and culture; this is an answer to one of the deepest desire of the human spirit" (MP 84). The Pope after his introduction further addresses the public about the duties of the listening public, which are vital in making radio broadcast serve its purpose, especially if it is a Catholic medium.

Radio is a listening medium since everything that is broadcast incorporates sounds via airwaves; it is different from cinema or television, which combines both sound and vision. Whatever is transmitted must not be taken for granted. Hence, the Pope emphasises that "the first duty of the listener to the radio is to make careful and deliberate choice from among the programmes offered. These must not be allowed to enter the home indiscriminately. They must be admitted, as a friend is admitted, on prudent and deliberate invitation" (MP 88).

## Religious Broadcast

The Pope did not just address the question of radio to the general or contemporary understanding and role of the media in society. He also spoke about the reality that even a medium such as radio must be used in the context of religion, for that is the core and fundamental principle when such media or any other medium is founded or established in the Catholic Church. He thus wrote that "by their rigorous exertions they have

[4] These words were an address by the Pope Cf. Broadcast message to those taking part in the third general convention for the study of national and international systems of communication, held on the 60th anniversary of the invention of wireless telegraphy, at Genoa; A.A.S., Vol. Xlvii, 1995.

achieved for religious broadcasting a position which befits the dignity of religion as more important than all other human affairs" (MP 95). In short, the Pope writing in this letter asks that all involved in running of these media—radio in particular—aim to promote the truth. This is in light of considering that the media in this day and age has become the servant of distorting the truth for convenience and profit over respect for human dignity. This is the essence of what it means to have a Catholic or any religious radio station—always to promote the truth in all circumstances.

The preceding sections have emphasised the importance that the Catholic Church attaches to all forms of media in its aspect of evangelisation. The Church knew that when these forms of evangelisation were being developed, and it had to stay on top to enhance the role of evangelisation. At various times the Church has addressed the importance of media, to the point of proclaiming a World Communication Day that falls generally around May, annually.[5] To make the Church's multiple apostolates in social communication more effective, a day is to be set aside each year in every diocese, at the bishop's discretion, on which the faithful will be reminded of their duties in this domain. They should be asked to pray for the success of the Church's apostolate in this field and to contribute towards it, their contributions to be scrupulously employed for the support and the further development of the projects which the Church has initiated in view of the needs of the entire Church (IM 18).

## The Story of Radio Veritas

This section looks at the contribution of Radio Veritas as the only Catholic radio station in South Africa. I tell this story from my perspective as someone with a close eye on how the process unfolded, as well as someone who has had experience setting up a Catholic radio station in another context. Radio Veritas' story is one of resilience, hard work and, above all, the faith and optimism of its founder, Father Emil Blaser, and a few committed and dedicated staff members. They never gave up trying to get the station on the air and keeping it there over the years since it was launched in 1999.

[5] It is important to point out here that since 1967 (during the reign of Pope Paul VI) every pope has released a World Communication Day letter that is officially published on 24 January, each year, the day we commemorate Saint Francis De Sales who was known for prodigious writing and was declared a patron saint for all journalists.

This is a unique feature, as indicated earlier, that most countries in Sub-Saharan Africa have several radio stations owned by respective bishops in the dioceses or sometimes owned by religious congregations, like Oblate Radio *Liseli* (Light) in Zambia where I established this radio station for the Congregation of the Missionary Oblates of Mary Immaculate. On the other hand, I was taken aback on visits to South Africa that there was only one Catholic Radio station compared to many countries even in the Southern African Development Community and Common Market of Eastern and Southern Africa region. However, it should be noted that attempts were made to set up a radio station in the Diocese of Upington long before the conception of Radio Veritas. A lay member (non-priest or non-religious employee) was assigned to investigate the prospects of setting up such a medium. However, things did not materialise for some reason, so the dream was over.

Years later, in came the Order of Preachers (OP) also known as the Dominicans (named after their founder, St Dominics). It was in 1999 that the first and only Catholic radio station (Radio Veritas) was founded and handed over by the South African Catholic Bishops' Conference (SACBC) to Father Emil Blazer, OP (Order of Preachers Priest). At that time, he served as the Media Officer of the SACBC. Though it was not easy to obtain the licence, his patience and perseverance paid off; a licence was granted to broadcast on medium wave and DSTV radio channels. It was felt that while it was essential to have churches and various centres like retreat centres for evangelisation and social influence, the means of communication was lacking, and radio was seen as a critical tool for integral evangelisation (which combines spirituality and social concerns). Here is a journey of how Radio Veritas found itself where it is as we commemorate 100 years of radio in South Africa. The radio station is currently under the direction of the Station Manager Khanya Litabe[6] who has been with it since its inception.

Radio Veritas was initiated by the South African Catholic Bishops' Conference in 1998 while our Director Fr Emil Blaser was employed there as Associate Secretary General and Media Officer. The Radio Veritas Trust and Board of Trustees were formed, and a constitution produced. Radio Veritas made the first of many applications to the Independent

[6] Khanya Litabe has been at the helm of Radio Veritas with Fr Emile Blaser, OP (RIP). Most of the information on Radio Veritas has been provided by him as well as the vision of the radio station.

Communications Authority of South Africa (ICASA) in 1998 for a medium wave community radio licence. The genesis of it all was that in 1999 the SACBC decided to hand over the project to the Dominicans, who felt that in that early stage only a production house would be feasible. Office space and a studio were hired and after six months the operation was moved to the Dominican property in Johannesburg. Programmes were produced and sent to Christian community radio stations around the country and were also made available to Catholics through bookshops and parishes. Despite these efforts being made, the style encountered reservations from non-Catholic radio stations in broadcasting material that came from a Catholic production house.

The challenge was the protracted years of licence applications, deputations, hearings and meetings with the Independent Communication Authority of South Africa (ICASA). ICASA, like its counterparts in many other countries, is the sole regulatory authority in terms of setting up broadcasting facilities. During this time of impasse, the Radio Veritas team used various initiatives to do some broadcasting, such as using the platform for two hours a day on a Portuguese medium wave frequency, Radio Cidade, and three to four times a year on one-month special events licences. The radio station was also broadcasting short wave and the DSTV satellite service. Later on, a two-hour Sunday spot on the National Broadcasters' Radio 2000 frequency was secured and utilised.

Meanwhile during this wait most staff members—volunteers and young people who had no background in radio—were being trained and honing their skills as broadcasters, news readers and technical producers. The team is proud and grateful for the opportunity to have played a role in the skills development of young Catholics in South Africa, since 1998. Many have taken managerial positions in commercial stations and the national broadcaster. Some have continued to work for Radio Veritas and are its backbone—one of them is Khanya Litabe, the Station Manager. In July 2006, Radio Veritas bought property and studios were built through the generosity of several Catholic donors. It was thus able to move into its present facility. The building is jointly owned by Radio Veritas and the Knights of da Gama.

Finally, after 12 years, the station was granted a licence in late 2011 and the first official broadcast as Radio Veritas 576 AM was Holy Mass, heralded by the joyful Easter Sunday bells at 9 am on 8 April 2012.

Every media outlet must have a vision for its programming so that people may understand ethos, including the producers themselves as to what it entails to be on air on a Catholic radio station. Therefore, Radio Veritas set up this vision:

- Our vision at Radio Veritas is that through our programming we can highlight and help people find areas where they can serve and become involved in the community.
- We believe firmly in the concept 'I am my brothers'/sisters' keeper.' We believe the role of Radio Veritas is not only to talk about the Faith, the Scriptures, and air Devotional Programmes.
- We believe we have a very powerful and prophetic challenge, through our programming, to highlight the needs in our country, and call one another to live the Gospel values of love, compassion and service, in active and practical ways.
- We believe that our programming must not only educate, inform, and entertain, but it must also inspire people to become who God created them to be.

The radio broadcasts in four African languages and in English and Portuguese. Its programme schedule consists of devotions (daily on-air Mass, rosary, angelus, stations of the cross), programmes on scripture, catechetics, phone-in shows, magazine programmes, current affairs, and youth, children's and family shows. In addition, the station includes news bulletins and interviews and work in tandem with several church organisations. It has also initiated a number of community initiatives, like collecting food and water to donate to areas during droughts. Until the 2019 COVID pandemic Mass was aired from different parishes every month, and events such as ordinations of bishops and priests, funerals of prelates and priests, and religious professions or anniversaries were broadcast.

The radio station has remained committed to being involved in the community and has trained many volunteers to present programmes and started a "Catholic School of Broadcasting", which had been put on hold since the Covid pandemic. They are now slowly getting back to this vision. The founder of the radio station, Fr Emile Blaser, passed away in 2020[7],

[7] Khanya Litabe gave this tribute at his passing: "In 2020 we lost our beloved founder Fr Emil Blaser, who died after a long illness which had left him unable to come to the office for 2 years, but did not dim his passion for and commitment to serving the Lord and the Church through Radio Veritas. In tribute to his work and his faith, we will continue to work to make the station what he dreamt it would be—an inspiration to all who long for a relationship with the Lord".

having achieved his dream of having a Catholic radio station on air in South Africa almost a century after radio broadcasting was first introduced.

## Radio Veritas' Current Contribution to the South African Society

The presence of Radio Veritas has added value to South African society, especially as far as religious broadcasting is concerned. Usually, religious programming is done part-time by buying airtime to broadcast messages on other radio stations. Owning and operating a Catholic radio station in South Africa is unique to Radio Veritas. In *Radio, Public Life and Citizen Deliberation in South Africa*, edited by Sarah Chiumbu and Gilbert Motsaathebe, several observations are highlighted about the importance and contribution of Radio Veritas. The radio station has had a unique way of making the audience participate in its programming: "our research has revealed that radio Veritas as a community radio station encourages audience participation with phone-in programmes and social media and plenty of phone-in radio shows on air" (Karombo & Kibombwe, 2020, p. 146). Further, it was discovered and concluded that:

> What differentiates this genre from those in the commercial sector, and public radio stations, is the amount of space given to listeners to air their views. The programme *As the Sun Set* broadcast on February 15, 2020, serves as a good example of audience participation in programming. In this radio show, the discussions with each listener lasted three to four minutes. The listeners participated using SMS, and phone calls during this broadcast. (Karombo & Kibombwe 2020, pp. 146–147)

Although Radio Veritas is a Catholic radio station, one cannot but notice that its aim is to be a real community radio that interacts with the people and makes the audience the main stakeholders in this media regardless of their background. Most community radio stations, even those that are non-religious, promote their audience's participation to remain consistent with their ethos. Radio Veritas in this regard is not exempt from these expectations. With the coming to the fore of social media, it is expected that such a medium provides the opportunity for even greater general public and audience participation. In earlier research, it was concluded that Radio Veritas has a large following due to its programming that has religious content but includes other aspects such as education, culture and various interest groups, to mention a few. The research findings highlighted that:

> *Radio Veritas* plays a critical role in the South African radio public sphere. The station has made massive strides in being inclusive, serving minority and marginalised groups that are alienated by the main stream media, such as older people, children and the poor. (Karombo & Kibombwe, 2020, p. 151)

With its large following by a Catholic community and other religious groups, the radio contributes significantly to the community's development as far as engaging the audience is concerned. There is enough evidence of goodwill from the listeners, well-wishers and the public to warrant the continuation of the station as indeed the people's voice from the community. The contribution of the volunteer base is vital to the radio station as they help to minimise the costs of production and running the radio station. The importance of volunteers is critical as "volunteer participation is readily available from the communities; without this type of engagement from the broadcast communities, these stations would not be able to survive" (Karombo & Kibombwe, 2020, p. 151).

Within this context of celebrating 100 years of radio in South Africa, Radio Veritas has built up a greater audience of listeners. Listeners who are quite involved in programming or participating in various phone-in programmes that deal with religion, health, social justice, education, culture and other aspects of society's needs.

## Conclusion

This chapter provides a broad understanding of the importance of media in South Africa as the nation commemorates 100 years of radio broadcasting. The study looked at a brief history of radio broadcasting in general and in the context of South Africa. The chapter has highlighted the Catholic Church's understanding of media in general (focusing on the five key fundamental documents or corpus of Catholic media). From the vast presentation of the critical key documents of the Catholic writings on the media, the chapter focused on Pope Pius XII's ***document Miranda Prorsus***, with particular emphasis on the radio section of this document. This was the first document in the history of the Catholic Church and its media to focus on Cinema, Radio and Television. The other aspect that this chapter looked at was to understand what Radio Veritas is, its historical context and what impact it has had in the South African context as a religious radio that has contributed within the 100 years of radio to the community of faith and society in general.

## References

Karombo, S., & Kibombwe, F. (2020). Catholic Radio and Its Importance in South Africa: A Case Study of Radio Veritas. In S. Chiumbu & G. Motsaathebe (Eds.), *Radio, Public Life and Citizen Deliberation in South Africa*. Oxon, UK: Abingdon.

Osunkunle, O. O., & Wozniak, J. (2015). Faith-based Radio: Beyond Spirituality, Encompassing Development. *African Journalism Studies, 36*(2), 77–94. https://doi.org/10.1080/23743670.2015.1043692

Pope Pius XII. (1957). *Miranda Prorsus (The Cinema, Sound Broadcasting and Television): Encyclical Letter of Pope Pius XII*. Catholic Truth Society.

CHAPTER 6

# More Than Just Edutainment: Indigenous South African Community Radio Music Programmes as Preservers of African Traditional Cultures

*Tshepang Molale* and *Quatro Mgogo*

## Introduction

According to the Broadcasting Research Council, as of 2021, there were 273 community radio stations across South Africa that service millions of listeners daily (BRC, 2021). This demonstrates the ubiquitous nature of community radio in the country. However, a careful review of the literature revealed that existing studies on community radio tend to focus on themes such as language preservation, mainly because they are broadcast

T. Molale (✉)
Department of Media Studies, University of Witwatersrand, Johannesburg, South Africa

Q. Mgogo
North-West University, Mahikeng, South Africa
e-mail: Quatro.Mgogo@nwu.ac.za



S. Nkoala, G. Motsaathebe (eds.), *100 Years of Radio in South Africa, Volume 2*, https://doi.org/10.1007/978-3-031-40706-2_6

in specified regional languages (see Mabika & Salawu, 2014), development and empowerment of community members from a participatory communication perspective (see Leketanyane et al., 2021), the marginalisation of indigenous languages at the behest of the English language that serves as lingua franca (Mohammed, 2019), talk radio and citizen participation from a social responsibility theoretical perspective (Nkoala & Matsilele, 2021), implications of gatekeeping and the Habermasian concept of the public sphere (Motsaathebe & Tsarwe, 2021), or the use of indigenous language radio in facilitating conflict resolution (Oburu, 2016).

Although these themes are essential in demonstrating the struggle fought by community media towards community development and the advancement of quality of life for the subaltern, there is a dearth of research interest devoted to the role played by community radio's traditional music programmes in protecting African traditional cultures against extinction. This chapter aligns itself with Molefi Kete Asante's for a "return or re-memory of the nature of African communication within the context of tradition, community, and value" (Asante, 2018) after carefully observing a moral crisis faced by the field of communication in Africa as a result of Western imperialism and domination.

The chapter begins by providing an overview of the two stations used in this multi-case study to contextualise community broadcasting in South Africa broadly. It then turns its attention to the dominant themes, theories and perspectives that inform scholarship on community media in this context. This is followed by a section on the study's theoretical points of departure along the lines of decoloniality and the role of African traditional music programmes in cultural preservation. This section also reviews community radio and traditional South African music programming literature. As part of this, it considers the significance of music as a tool for African renaissance, tracing this back to when radio was first introduced to black audiences in South Africa in the 1940s.

## Inkonjane FM and Mahikeng FM: An Overview

Inkonjane FM and Mahikeng FM are two indigenous language community radio stations set up as part of efforts to transform the country's media and broadcasting landscape. This was to protect indigenous languages in an industry dominated by English language hegemony, enable active citizen participation in the production and dissemination of local content, and allow local citizens to also have a say in issues affecting them, either from a national or local context. These stations were also aimed at bringing government information around issues of development and social change closer to the communities that otherwise are far from reach, and providing a platform for negotiating identity and belonging while undoing the racial injustices of the past where economic gains in the media and broadcasting sector were only shared among a few elites.

Inkonjane FM's name draws inspiration from the swallow-like bird called *inkonjansa* in the IsiXhosa and IsiMpondo languages. This is because the swallow can fly so high and cover a long distance within a short period. Thus, Inkonjane FM is envisioned to behave the same using sound waves. Its slogan, *sixoxa sonke*, directly translates to "we all talk" to encourage the community to be part of the community developmental talk shows. It is located in Flagstaff, a rural town within the OR Tambo District Municipality, Eastern Cape Province, South Africa. Its coverage transcends the Pondoland area. Even though the station was established on 7 July 2006, and licensed by the Independent Communications Authority of South Africa (ICASA) on 24 October 2008, Inkonjane FM began its broadcast on 20 April 2009. It is split between 60% talk and 40% music, where the community can engage in discussions while at intervals enjoying music, including the traditional music of AmaMpondo.

AmaMpondo, a subculture of AmaXhosa (Mgogo & Osunkunle, 2020), results in Inkonjane FM broadcasting mainly in the isiMpondo language, mostly a combination of the IsiXhosa and Zulu languages (Reid & Mckinley, 2020). To accommodate other cultures within Pondoland and beyond, it broadcasts 70% in IsiMpondo, 20% in IsiXhosa and 10% in English. The station has a listenership of 120,000 and is currently ranked number 2 in the Eastern Cape Province based on its popularity. It has two traditional African music programmes on the Mpondo cultural tradition. This Mpondo traditional music is called *Ingcambu zethu*, which directly translates to "our roots". The second one is called *Sidudli'fene*, characterised by Maskhandi music (involving Xhosa and Zulu traditional music). However, the study focuses on *Ingcambu zethu* traditional music programming since it is mainly about the Mpondo cultural tradition.

*Ingcambu zethu* is 70% Mpondo music and 30% talk about traditional and cultural issues while educating AmaMpondo about their culture and customs to maintain the values of the ethnic group. In this programme, Mpondo traditional leaders and Mpondo elders participate in helping to preserve the tradition and culture of AmaMpondo and advise on the correct way of performing customs and traditions of AmaMpondo. Similarly, the station's mission is to become the progressive soul of the AmaMpondo for community development and inspiration while educating the community about the inspiration embedded in the culture in a manner the community understands and can appreciate.

The second station considered is Mahikeng FM. As its name suggests, Mahikeng FM is located in the capital city of the North West Province, Mahikeng, South Africa. Although it was established in 2005, the station began operating in 2008 after receiving a community broadcasting licence from ICASA. In terms of its licence conditions, Mahikeng FM is a 60% talk and 40% music station which aims to "enrich", unify and create a platform where the "community can fully express itself" (Kgokong, 2022). It mainly broadcasts in Setswana and serves communities in and around Mahikeng, including rural villages such as Magogoe Village, Disaneng Village, Lotlhakane and Bodibe. Communities in these villages surrounding Mahikeng are predominantly Setswana-speaking. The station boasts a listenership base of about 124,000 listeners and has received numerous awards over the years as one of the leading community radio stations in the North West Province. It has one African traditional music and talk programme named *Phafana ya Setso.* This name has a rich cultural meaning in Setswana culture and tradition; the word *Phafana* can be loosely translated to mean a "jar that men use to pour and drink traditional African beer" during their meetings, also called *Kgotla*, while the words "*ya Setso*" can be directly translated to mean "of culture". Therefore, *Phafana ya Setso* means a jar from which we can drink and gain cultural knowledge. This show and its meaning are relevant because several subcultures reside in Mahikeng and surrounding villages; we have Batswana and iSiXhosa-speaking communities and Zimbabweans who are Shona-Speaking. The show was previously called *Phafana ya Setswana* (meaning a jar from which we can drink/gain Setsawana cultural knowledge). However, it was changed to its present name because of the abovementioned different subcultures in Mahikeng. The station aims to use this radio programme to unify and teach community members about the different cultures found in their communities (Medupi, 2022).

## A Decolonial Turn Towards African Language Community Radio

In considering these specific stations, and community radio in South Africa broadly, the view is that this segment of the radio sector is one of the means through which communities have attempted to use media to address the problematic features of a 100-year-old industry that primarily existed to serve the colonial and apartheid agenda. Tyali (2021) asserts that community radio in South Africa is decolonial in that it promotes African interest broadcasting through a cultural "liberatory" role it plays "in relation to the history and memory of a particularised African community" (Tyali, 2021, p. 192). Furthermore, through broadcasting content such as indigenous African music, community radio espouses a decolonial posture by serving as spaces for memory, thereby demonstrating a unique "potential to anchor and connect current generations to their history and memory" (Tyali, 2021, p. 195).

In considering how indigenous African music programmes aired on community radio help to facilitate local culture and everyday life, this study draws on Mano and Milton's (2021) critical discussion on media, Milton's communication and decoloniality, and community media. The view is to challenge the marginalised South Africans. Radio's place has primarily been educating, informing and entertaining. Case studies show how cultural preservation becomes an important decolonial descriptor for community radio within a post-apartheid South African context. In this way, the "colonial" legacy of radio in South Africa, particularly in terms of English language hegemony (see Motsaathebe, 2010), is subverted in that community radio serves to create bottom-up approaches, where local communities are at the forefront of content creation (through bottom-up citizen participation), thereby emphasising indigenous language use (see Leketanyane et al., 2021).

## Literature Review

### *The Significance of Indigenous African Music as an African Renaissance Tool*

In the post-colonial era, characterising indigenous African music is becoming a daunting task. This is mainly because of the erosion of indigenous African music or the Western influence from both the colonial and postcolonial eras. However, if we are to characterise it from its purity, as Mugovhani (2015) argued, indigenous African music is mainly associated with traditional ceremonies, rituals, entertainment and social cohesion. Furthermore, indigenous African music is usually aligned with unique costumes and is accompanied by culturally defined rhythms.

Noting that indigenous African music is highly improvised (Lawn, 2013), a principal rhythm is played using a drum made of wood and cow skin. Then a song characterised by call-and-response will be passed down orally among the audience or backup singers. Then, the traditional dance moves harmoniously to the rhythmic patterns performed. Borrowing from Mkhombo's idea of indigenous African music as a renaissance tool, it can also be argued that indigenous African music symbolises African culture. More so, it reinforces social cohesion, educates on the nation's history, communicates socio-economic aspects, restores traditional values of the society, and promotes the importance of Ubuntu (Mkhombo, 2019).

However, the influence of Western popular culture has arguably affected current generations such that it might as well be the case that indigenous African music, language and culture are facing extinction. In response to this concern, it is believed that "indigenous African music restores the status of indigenous African languages" Mkhombo (2019, p. 168), and community radio plays an integral part in facilitating this restoration. Considering this view, community radio can be a tool in promoting the African renaissance. Koma (2018) defines the African renaissance as the "rebirth, revival and reawakening" of the African continent with self-identity, critical awareness and self-determination as its core values. Furthermore, the African renaissance can be understood as the revival of what has been lost in culture, tradition and African philosophy (Mekoa, 2018).

Diniso (2017, p. 124) notes that the "African renaissance is closely related to *ubuntu*". Considering this view, it is noted that indigenous African music can be used as a tool for driving African renaissance ideology. Thus, we note that indigenous language music programmes aired on Inkonjane and Mahikeng FM community radio stations are not only the preservers of African traditional cultures but also serve as a revival of aspects of our traditions and cultural values that were eroded as a result of colonialism and apartheid. Thus, this view can be understood to serve the idea of the African renaissance. Borrowing from Motsaathebe's (2010) critical reflection on language discourse in Africa, we argue that the community radio sector in South Africa plays a liberatory role which helps advance the African renaissance, thereby promoting Tyali's (2021) "African interest broadcasting" in Tyali's *An Afrikological Perspective* (Motsaathebe, 2011). Here, Afrikology is understood as deliberate intentions to promote African thinking, cultures, values and ways of life as part of promoting Africanisation along the lines of decolonisation (see Mano & Milton, 2021).

### *Indigenous Language Media: Language, Identity and Cultural Preservation*

A careful review of the literature demonstrates a sustained but nascent effort from parts of the African continent, particularly in the Southern African Development Community, devoted to exploring the nexus between African traditional/Indigenous language media, language, identity, and cultural preservation. Although much research interest is centred around the role of community radio in the promotion of development and social change (see Chikaipa & Gunde, 2021), critical studies can be found where the status quo is explored insofar as the use, appreciation and marginalisation of African indigenous languages in the media public sphere.

Often, when scholars explore the status and role of indigenous languages, it is a given that they mainly focus on themes such as the political economy of indigenous language media, the use of indigenous languages on digital platforms, and innovative ways in which technology-mediated communication can help in the preservation of indigenous languages, as well as the practice of indigenous language journalism on online platforms, strategic ways in which indigenous language media can sustain themselves, and new research agenda and emerging trends for indigenous language media in Africa (Fadipe & Salawu, 2022a; Mpofu & Salawu, 2020; Oyesomi & Salawu, 2020; Salawu, 2019; Salawu, 2021a; Salawu, 2021b; Tshabangu & Salawu, 2022a; Tshabangu & Salawu, 2022b). Unfortunately, research groundwork is predominantly narrow and underexplored in terms of indigenous language music, community radio and cultural preservation. The nascent body of emerging work regarding indigenous language music and its role in advancing culture and ethnicity, as well as regional identities, is attracting research attention (cf. Fadipe & Salawu, 2022a, 2022b; Fadipe & Bakenne, 2022). However, the gap is in the use of indigenous language community radio and its music programmes towards advancing and preserving cultural practices—a gap the present chapter aims to fill.

### *Edutainment, Community Radio, Traditional Music Programmes and Cultural Preservation*

The literature is well populated with various case studies and insights on the importance of community radio in promoting citizen participation, development and social change. Within the South African context, various studies have been conducted on this subject matter, but they rarely address the issue of using community radio's indigenous and cultural music programmes to serve local cultures. Themes that are primarily discussed in the literature within an African context include community radio and citizen participation (Leketanyane et al., 2021), community radio and the democratisation process, popular music and political communication (Fadipe, 2016), and regulation and policy development for community radio programming (Banda, 2003).

Insofar as music and identity are concerned, an ethnographic study by Bodunrin (2019) explored how the youth from !Xun and Khwe communities in Platfontein, near Kimberly in the Northern Cape, South Africa, "negotiate a balance between the dominant western popular culture and their own indigenous culture by effectively using their indigenous languages in a hip-hop-driven modernity" (2019, p. 175). As the author further postulates, the benefits of such negotiation are found in the oral styles of hip-hop music which enable the youth in Platfontein to "creatively incorporate and transform their traditional practice of storytelling using modern digital recording technologies" (ibid.). This is one of the few critical studies dealing with decoloniality, indigenous language preservation, and cultural and identity preservation in South African radio over the past 100 years.

## Methodology

This chapter adopted a qualitative research methodological approach and a case study design. In this way, perceptions of the different role players, such as those in the management of the two community radio stations and listeners, were obtained through interviews. The study purposely selected Inkonjane FM and Mahikeng FM community radio stations as case studies. This case study selection was based on the fact that both radio stations are community-driven, they are indigenous language community radio stations, they are popular in the areas in which they operate, and the specified shows are cultural niche programmes that resonate in the communities in which they broadcast.

The data were collected through key-informant interviews with four community members in Mahikeng, the capital city of the North West Province, and eight from eMampondweni (Flagstaff), a rural town in the OR Tambo District Municipality. Using this method, four indigenous language music programme producers and station managers were interviewed at both community radio stations. Thus, a total of N=13 respondents participated in this study.

First, fan club members of the station were identified using social media. Radio station fan clubs are supporter groups which comprise regular listeners. These groups are often formed on social networking sites, such as Facebook. Researchers then identified one fan club member of each radio station, and they were each requested to identify those that they know are also members of these fan clubs and avid listeners. Finally, each respondent was asked to recommend another fan club member, and the data was collected until the data saturation point. For in-depth interviews, station managers and presenters at both stations also fit the criteria and were thus identified by purposive sampling.

The data were analysed using the thematic analysis method. Given the focus of the study, two themes were considered: cultural preservation and community or citizen participation. Cultural preservation is related to discussions about, for example, the construction and promotion of regional identities and the accommodation of cultural diversity. At the same time, community and citizen participation are related to the role played by communities in the sourcing, production and dissemination value chain of indigenous language music content. The following section presents the study's findings and a synthesised discussion of the data.

## Discussion

### *Cultural Preservation Through Community Radio's Indigenous Language Music Programmes*

Radio's defining culture has transformed from being difficult to extremely political (Harding & Cortes, 2018). Culture refers to uniform patterns of behaviour deeply rooted in a unified society (cf. Abdussalam et al., 2021). However, we also note that culture and language are inextricably intertwined. According to Chen and Le (2018), culture and language are two sides of the same coin, or culture is language, and language is culture. Therefore, using indigenous language in the media (e.g. community radio) plays a crucial role in asserting identity and language status (Motsaathebe, 2018).

Similar to the main argument of this chapter, such media as community radio plays a significant role in preserving culture and language and promoting the community's status and culture and tradition. Over the 100 years of radio in South Africa, even in periods when the content aimed at indigenous language speakers was of a propaganda nature (Mhlambi, 2015), the simple feature of being able to hear one's language on air disrupted the prevailing norm of excluding African cultural expressions on air.

The establishment of community media, radio in particular, must be read through this lens, being aware that one of the impetuses that gave rise to the formation of this radio segment was a need for platforms that would allow diverse communities to express themselves linguistically and culturally. One of the core functions served by community radio is to present community interests and enable maximum community participation. In light of these views, respondents were quizzed about the ability of the community radio's indigenous language music programmes to preserve the radio's culture and the language of the community it serves. Several standard views from the key informants were noted on both radio stations. Below are some of the opinions that came up during the in-depth interviews at Inkonjane FM in response to the question:

> Yes, they do. The program teaches other nations the culture of Amampondo, and they encourage Amampondo to think positively about their heritage and not to think that their way of singing is unacceptable. (Inkonjane FM presenter)
>
> Yes, it does preserve the culture. However, as Nelson Mandela once said, you must speak to the people in the language they understand. We are doing just that at Inkonjane FM. Because the language of IsiMpondo is not just language. It carries the culture and heritage of Amampondo. And as a station manager, I must ensure that I filter and verify every piece of content coming into the station to ensure that it is mixed content but strictly verified content that represents Amampondo. (Inkonjane FM station manager)
>
> Yes, they do. When I listen to programs like Siudlimfene, ingcambu zethu, and ezidlumzi on Inkonjane FM, I always feel so proud to be iMpondo. I also feel that our heritage is safe. (Inkonjane FM Listener 3)

For Mahikeng FM, similar views were noted, and they have been extracted as follows:

> Yes, the Setswana traditional music programme has done a lot to preserve Setswana cultural music in Mahikeng; for example, I grew up listening to another Setswana radio station, and I do not remember it having a dedicated show for cultural music for Batswana, just like Mahikeng FM; therefore I think this programme about Setswana music and culture has made a big difference, especially on the youth, here, in Mahikeng. (Mahikeng FM Listener 3)

Another Mahikeng FM listener shared a similar view and shared personal experience of how their life was affected by the indigenous traditional music programme:

> I have learnt that culture is cultural; you cannot ignore it even if you can adopt other cultures, however somewhere, you will have to go back and revisit your culture; for example, if you have a partner, you will have to pay their lobola so that your children can have your surname. If you do not do so, then your children will not carry your surname. (Mahikeng FM Listener 4)

In demonstrating their personal experience of how the indigenous language music programme has played a role in their life, the listener further made an interesting remark about a lesson they learned from one of the songs by a prominent Setswana traditional music artist, Mmaausi, whose music is regularly played on *Phafana Ya Setso*:

> Mmaausi asks: 'did you inform the uncles that their nephew is getting married? No wedding can go on without inviting uncles and their wives to play a role in the process'. (Mahikeng FM Listener 4)

In the above interview extracts, it was observed that cultural preservation is an important personal factor for the listeners at both radio stations. As such, having traditional indigenous language music programmes on the community radio stations promotes cultural assertiveness and pride for local indigenous communities. However, it cannot be said that these community radio stations successfully promote wide-spread cultural preservation for the Batswana and amaMpondo ethnic groups. This is predicated on the fact that, for example, in comparing the listenership statistics of

Mahikeng FM against the latest census in Mahikeng, only about 39.4% of the population listen to Mahikeng FM (i.e. 124,000 listeners out of a population of 314,319). To arrive at this conclusion, the latest listenership statistics provided by Mahikeng FM's CEO were subtracted from official population statistics extracted from the 2016 Community Survey indexed by Media Monitoring Africa's Map. In the case of Inkonjane FM, it was discovered that only about 6% of the amaMpondo community listens to the radio station. This is based on the fact that the station has about 120,000 listeners out of an estimated two million amaMpondo community members, as evidenced in the *Encyclopedia of African Peoples* (The Diagram Group, 2013).

This is a striking discovery for both stations. It confirms arguments and observations made by various scholars, such as Molale and Mpofu (2021), Salawu (2021a), as well as Mpofu and Salawu (2018), on the lack of popularity of indigenous languages in the media space owing to, among many factors, apathy demonstrated by the general public towards learning and consuming indigenous language content in the media, as well as the existing English language hegemony that is threatening the survival of indigenous languages across the mediascape.

Furthermore, although it was found that the indigenous language traditional music in community radio has a role to play as a resilient communication medium for language revitalisation (cf. Danos & Turin, 2021), there are some listeners whose cultures are not represented in indigenous language music programmes. Therefore, they tend to feel like outcasts in the communities within which they live. For instance, one listener to Mahikeng FM shared the following sentiments:

> I, as a Xhosa-speaking person, who was born in Mahikeng, my culture and tradition do not feature, and, as such, they are not addressed… It is painful; it is like we have been isolated in the land of Baroit's, and what if I was born where my forefathers were born in Eastern Cape, but now I have no choice, Sir, I have to accept the status quo. (Mahikeng FM Listener 3)

This sentiment contrasts with the reason Mahikeng FM's CEO provided when explaining why the name of their indigenous language traditional music programme was changed from *Phafana ya Setswana* to *Phafana Ya Setso*. A disconnect is picked up because, on the one hand, the CEO stated that the programme was changed from promoting Setswana's culture to accommodating diverse cultures found in Mahikeng, including

Setswana. On the other hand, the Xhosa-speaking community member argues that their culture does not feature in the programme. A yearning was sensed for Xhosa traditional customs and traditions to find a connection to the radio station, mainly because the listener is an avid fan of the station. This yearning can be demonstrated in the following interview extract:

> I have never heard any Xhosa traditional songs; apart from that, there is nothing that I can say. What I like is this traditional music show reminds young people about where the Batswana come from and what their tradition is all about. (Mahikeng FM Listener 3)

In light of the foregoing comments, it is clear that an opportunity exists for indigenous language community radio to become a tool for creating and promoting regional and national identities (cf. Chari, 2021; Mpofu & Salawu, 2018). In addition, community radio stations also need to make a more deliberate attempt to foster and promote content to accommodate a multiplicity of cultures. The regional population makeup of communities across South Africa constitutes a diversity of cultures and multilingual societies.

### *Community and Citizen Participation*

Community radio is known for including the community it represents in the development programmes. It is known for allowing the community, who are the listeners, to feel that they are part of their personal development (see Osunkunle, 2013). This is synonymous with the view that community radio, for instance Inkonjane FM and Mahikeng FM, should be the centre for community participation and development, especially in the indigenous language music programmes, which are perceived as community development in terms of cultural, traditional and heritage preservation and promotion. This view is also similar to Mabingo's (2020) argument that indigenous dance and music not only preserve the culture of the community but are also community developmental. Mabingo further notes that the contextualisation and adaptation of indigenous dances and music boosts indigenous knowledge and skill relevance, viability and applicability.

Based on the empirical data mined from the participants of this study, it was found that, to a great extent, the radio station does allow the community to be part of their development programmes, like indigenous language music programmes. However, the community does not reciprocate equally. For instance, based on this view, it emerged from one of the key informants from this study that the content of the indigenous language music programme is primarily sourced (recorded for radio) from the community cultural programmes that are ritual based. As a result, it was noted as part of community participation. However, on the other hand, it was mentioned in some of the cultural events, considered sacred, that the content is not allowed to be sourced or recorded, and that limits the depth of the programme in terms of the indigenous content. For instance, below are some of the views that emerged during the in-depth interviews:

> When these programs were introduced, some community members felt it was enough to participate in the program when its presenter would invite calls for participation. Instead, they would come to the studio to listen to these programs while dancing to the indigenous language music. Some would come wearing traditional outfits and instruments like the guitar to dance and sing. To us, it showed how much the community wants to be part of our programs that represent them. (Inkonjane FM station manager)

In the extract from the above interview, it is notable that the station manager at Inkonjane FM believes that "participation" means creating a platform for listeners to listen to the traditional music programme, not only through radio but also by allowing them to also come to the station, listen and dance to tunes played on air. However, in contrast to the above view, the indigenous language music presenter understands participation as being able to create or source content that diverse groups can enjoy. This observation stems from, for instance, the following interview extract:

> The challenges I encounter are, first, because of demarcation and social discrimination. I am sometimes not allowed in certain traditional events to record music. Sometimes when the event is spiritually inclined, people do not allow me to record music. Therefore, if people can be taught about the purpose of radio and its mandate, it will be easier for them to let me record music at their events. Furthermore, if society can promote the culture of amaMpondo, it would be easier for the radio to reach more audiences involving the chiefs and saps to resolve the social discrimination issue. It can also help. (Inkonjane FM presenter)

However, listeners understand participation from the vantage point of access to the indigenous language music programme during its broadcasts. For instance, one of the listeners mentioned the different platforms through which they access traditional music programmes:

> Yes, the radio station is open for listeners to participate in indigenous music programs like Sidudlimfene and ingcambu zethu. These shows also have online platforms like the Facebook page, where we can talk and comment during and after the show. For example, when Bhabhezilokhweni's Sidudlimfene show plays on Saturday, we comment on its Facebook page. We are also allowed to suggest traditional music artists and songs that are new to be played on the show. (Inkonjane FM Listener 1)

Similarly, in the context of Mahikeng FM, two divergent interpretations of citizen participation were noticed. On the one hand, some listeners and station members perceived participation in the same way as in Inkonjane FM, where it is viewed as mere involvement (i.e. participation as a means to an end). On the other hand, some interpreted participation from the vantage point of transformative discussions about content creation and decision-making around what content is needed (i.e. participation as an end). This is demonstrated in the following interview extracts presented in Table 6.1.

Considering the above views, participation is interpreted as "audience involvement". Here the research participants do not perceive the transformation character of participation. Perhaps, Cornwall's (2007) decade-old belief in participation, her complicated buzzword, is empirically confirmed. This finding was also observed by Leketanyane et al. (2021) in a study conducted at a selected community radio station. The observation made by the authors in that study is also confirmed. Participation is understood from a monologic sense—where content creators and producers hold the sole decision-making power to develop content and, in a linear form, transmit the content to the listener, who, in turn, is involved by responding to the content through enjoyment. On the contrary, participation is more than mere involvement of others; various recent studies confirm that participation is transformative in that it allows for dialogic communication where the aim is to enable all role players to make decisions about what they want, what they need, and how (Carpentier et al., 2019; Melkote & Steeves, 2015; Molale, 2019; Molale, 2021; Tau, 2013; Thebe, 2016; Tufte, 2017).

**Table 6.1** Divergent interpretations of "participation" (i.e. participation as a means to an end vs participation as an end)

| *Participation as involvement (participation as a means)* | *Participation as transformative dialogue and social change (participation as an end)* |
|---|---|
| We use different platforms in Phafana ya Setso; we "use our Facebook page called Mahikeng FM 96.7, which is our page. We then have WhatsApp, listener comments, and maybe voice notes or general comments. For example, we would have a topic we are discussing, maybe a proverb or idiom or short story, and we would then ask listeners to deconstruct or explain the different proverbs we have; then we would have you calling in to say I am so and so and I think the answer is in this or that way (Mahikeng FM presenter). | As a community radio station, we are mandated by the regulator via our broadcasting licence to get the community involved. Firstly, we annually recruit the general community to become formal members of the organisation, to enable them to have a say in the station's governance. We also host programming meetings (we have one this coming weekend), where the community is invited to come to give feedback on the current radio shows—are they satisfied with them, How can they be improved, what kind of subjects still need to be covered etc. [sic] (Mahikeng FM CEO). |
| Most of the time, as some shows come to an end, they would ask listeners to send messages or call to just express themselves on what was their experience of the show… listeners can of Mahikeng FM can improve the show by being involved in the activities of the station where most of the time as the station would conduct outreach campaigns, seeking comments from listeners/ community members on how to improve the different radio programmes. They should play an active role, and they should be present at such meetings; they should come up with ideas, different proposals and concepts about what programmes can be introduced that are relevant to their views, which can then be broadcast on Mahikeng FM (Mahikeng Listener 2). | To improve the participation of community members in Mahikeng, the station needs to give more opportunities to them. They should create more opportunities for our children because you can see that our children are easily enticed and are susceptible to peer pressure which ends up leading them to drug and substance abuse; it is now also in teenage pregnancy before they mature physically and mentally. So if they can be given more opportunities for participation, these social ills: drug abuse, teenage pregnancy could be alleviated, and our children can focus on cultural activities, and dance and they can play our cultural activities (Mahikeng Listener 1). |

In addition, the above interview extracts also raise awareness around nuanced interpretations of participation in development and social change: Should it be viewed from a top-down instrumental action perspective, where it is a means to an end, or should it instead be appropriated from a bottom-up vantage point where it is an end itself? Here the interview extracts of the CEO, as well as the listeners, show that they are aligned with literary calls on the issue as well as the perspective in the present study: If community radio is to be transformative, then participation (by the audience as citizens) should be approached as an end itself, not a means to achieve desired ends (cf., Andersson, 2017; Huesca, 2008; Johansson, 2020; Mefalopulos, 2008; Molale, 2019; Muturi & Mwangi, 2009).

## Conclusion

Our study has found that indigenous language music programmes in community radio stations can be tools for cultural preservation. Community radio's developmental and praxis-driven approach demonstrates an ability to go beyond merely functioning as a hub for edutainment as a normative requirement and expectation. Through active citizen participation in, for instance, outreach campaigns where meetings are held with listeners to inform community radio content and to restructure the different radio platforms, citizen participation in the stations becomes transformative in that the listeners become content creators and not merely mass consumers and recipients of the content. This demonstrates a complete shift in the conceptualisation of the role of indigenous language audiences from passive listeners, who were subjected to propaganda in the 1940s, to active participants who provide input on what is most relevant to them. Here, participation is deconstructed and appropriated instead of being understood as a means to achieve desired ends (e.g. holding community meetings to fulfil ICASA licence conditions). This is linked with accepted norms and standards in the literature about participation, as also argued in recent studies, such as Leketanyane et al. (2021) and Molale (2019, 2021).

Furthermore, the above role of community radio in promoting cultural preservation and in allowing active and bottom-up citizen participation in decision-making about content production can be construed as being decolonial in that it enhances the community's activeness in enriching African traditions through deeper participation in promoting indigenous language, music and indigenous knowledge. Therefore, Asante's (2018) call for a "return or re-memory of the nature of African communication within the context of tradition, community, and values" (Asante, 2018) is favourably answered. More so, it is challenging to delink this view from African renaissance's philosophy, as pointed out by Koma (2018). Therefore, as a result of these views, it is also concluded that the role of community radio transforms from preserving African traditional cultures and indigenous languages to becoming a tool for promoting the African renaissance. However, the study also discovered that more still needs to be done, at the practice level, to help educate listeners about what citizen "participation" entails, particularly in indigenous language music programmes.

Another issue that could be explored further is how participation is applied in community radio stations, mainly when decisions are taken about indigenous language content to accommodate local cultures. For practice, the study recommends that indigenous language community radio stations conduct community surveys more comprehensively to have a perspicacious understanding of the cultural makeup of their listener base. These surveys can help community members and the stations develop content inclusive of all diverse cultures and ethnic groups within their reach.

## References

Abdussalam, A. S., Aderibigbe, I., Babantunde, S. T., & Akindipe, O. (2021). *Culture and Development in Africa and Diaspora*. Routledge.

Andersson, E. (2017). Young People's Political Participation: A Public Pedagogy Challenge at the Municipal Level. *Young, 26*(2), 179–195. https://doi.org/10.1177/1103308817729971

Asante, M. K. (2018). The Classical African Concept of Maat and Human Communication. In K. Langmia (Ed.), *Black/Africana Communication Theory* (pp. 11–24). Palgrave Macmillan.

Banda, F. (2003). Community Radio Broadcasting in Zambia: A Policy Perspective. Unpublished doctoral thesis, University of South Africa, Pretoria.

Bodunrin, I. A. (2019). Hip-hop and Decolonized Practices of Language Digitization among the Contemporary !Xun and Khwe Indigenous Youth of South Africa. *Critical Arts, 33*(4–5), 174–190. https://doi.org/10.1080/02560046.2019.1702070

BRCSA: The Broadcast Research Council of South Africa. (2021). *July Overview*: BRC Interim Radio Data. Retrieved November 22, 2021, from https://brcsa.org.za/interim-radio-data-overview-overall-jan-june-2021/

Carpentier, N., De Melo, A., & Ribeiro, F. (2019). Rescuing Participation: A Critique on the Dark Side of Participation Concept. *Comunicação e Sociedade, 36*, 17–35. https://doi.org/10.17231/comsoc.36(2019).2341

Chari, T. (2021). News Syndication and Local Language Broadcasting in South Africa. In A. Salawu (Ed.), *African Language Media: Development, Economics and Management* (pp. 243–264). Routledge.

Chen, S., & Le, T. T. (2018). *Teaching Culture in English as an International Language: An Integrated Model.* Routledge.

Chikaipa, V., & Gunde, A. M. (2021). The Role of Community Radio in Promotion of Indigenous Minority Languages and Cultures in Malawi. *Journal of Radio & Audio Media, 28*(2), 327–343. https://doi.org/10.1080/19376529.2020.1751633

Cornwall, A. (2007). Buzzwords and Fuzzwords: Deconstructing Development Discourse. *Development in Practice, 17*(4–5), 471–484. https://doi.org/10.1080/09614520701469302

Danos, D., & Turin, M. (2021). Living Language, Resurgent Radio: A Survey of Indigenous Language Broadcasting Initiatives. *Language Documentation and Conservation, 15*, 75–152.

Diniso, P. M. (2017). The Role of African Theology and Spirituality in Bereavement Among Xhosa Children. PhD thesis, University of Fort Hare.

Fadipe, I. A. (2016). Political Communication and Indigenous Popular Music: The Study of *Apala* and *Fuji* Music Genres. *Journal of Communication and Media Research, 8*(1), 185–198.

Fadipe, I. A., & Bakenne, A. N. (2022). Extra-Mundane Communication in Ayinla Omowura's Music: Exploring Connections Between a Tool and an Agent. *Indigenous African Popular Music: Social Crusade and the Future, Vol. 2*. Palgrave Macmillan.

Fadipe, I. A., & Salawu, A. (2022a). *Strategic Communication: Can African Indigenous Popular Music Come to Play?* In H. Mangera, I. Mhute, & E. Jakaza (Eds.), *Strategic Communications in Africa: The Sub-Saharan Context.* Routledge.

Fadipe, I. A., & Salawu, A. (Eds.). (2022b). *Indigenous and Indigenised Yoruba Popular Music. Indigenous African Popular Music: Prophets and Philosophers, Vol. 1*. Palgrave Macmillan.

Harding, S., & Cortes, O. C. (2018). *The Routledge Handbook of Translation and Culture.* Routledge.

Huesca, R. (2008). Tracing the History of Participatory Communication Approaches to Development: A Critical Appraisal. In J. Servaes (Ed.), *Communication for Development and Social Change* (pp. 180–198). Sage.

Johansson, G. (2020). Participatory Art for Social Change? A Study of the Quest for Genuine Participation. MA dissertation, Malmo University, Malmo, Sweden.

Kgokong, K. (2022). A Manager's Perception on the Role of an Indigenous Language Community Radio in Preserving Local Cultures and Promoting Regional Identities Through a Traditional Music Show (*"Phafana Ya Setso"*), Mahikeng F.M. [WhatsApp/Social Media interview]. 17 March, Mmabatho.

Koma, S. B. (2018). The African Renaissance and the Impetus for Transforming Higher Education. *African Journal of Public Affairs, 10*(2), 97–108.

Lawn, R. J. (2013). *Experiencing Jazz: eBook Only* (2nd ed.). Routledge.

Leketanyane, P., Molale, T., Ogunsanya, A., & Asak, M. (2021). Deconstructing the Participation of Rural Dwellers in a Community Radio Station: A Participatory Development Communication Approach to a Radio Station in the North West Province, South Africa. In S. Chiumbu & G. Motsaathebe (Eds.), *Radio, Public Life, and Citizen Deliberation in South Africa* (pp. 50–66). Routledge.

Mabika, M., & Salawu, A. (2014). A Tale of Failure: Indigenous Language Radio Broadcasting in Zimbabwe. *Mediterranean Journal of Social Sciences, 5*(20), 2391–2401.

Mabingo, A. (2020). *Ubuntu as Dance Pedagogy in Uganda: Individuality, Community, and Inclusion in Teaching and Learning of Indigenous Dances.* Palgrave Macmillan.

Mano, W., & Milton, V. C. (2021). Decoloniality and the Push for African Media and Communication Studies: An Introduction. In W. Mano & V. C. Milton (Eds.), *Routledge Handbook of African Media and Communication Studies* (pp. 1–18). Routledge.

Medupi, T. (2022). Presenter's Perception on the Role of an Indigenous Presenter's Commmunity Radio in Preserving Local Cultures and Promoting Regional Identities Through a Traditional Music Show (*"Phafana Ya Setso"*), Mahikeng F.M. [WhatsApp/Social Media interview]. 14 March, Mmabatho.

Mefalopulos, P. (2008). *Development Communication Sourcebook: Broadening the Boundaries of Communication.* The International Bank for Reconstruction and Development/The World Bank.

Mekoa, I. (2018). Thabo Mbeki's 'African Renaissance': A Critique of Its Theory and Practice. *African Renaissance, 15*(1), 9–28.

Melkote, S. R., & Steeves, H. L. (2015). *Communication for Development: Theory and Practice for Empowerment and Social Change* (3rd ed.). Sage.

Mgogo, Q., & Osunkunle, O. (2020). Intercultural Communication Challenges and Its Effect on Students' Interpersonal Relationship at a South African University. *Journal of Human Ecology, 72*(1–3), 36–47. https://doi.org/10.31901/24566608.2020/72.1-3.3272

Mhlambi, T. N. (2015). Early Radio Broadcasting in South Africa: Culture, Modernity & Technology. PhD Thesis, University of Cape Town. https://open.uct.ac.za/bitstream/handle/11427/17260/thesis_hum_2015_mhlambi_thokozani_ndumiso.pdf?sequence=1

Mkhombo, S. M. (2019). The Status of Indigenous Music in the South African School Curriculum with Special Reference to isiZulu. Unpublished doctoral thesis, The University of South Africa.

Mohammed, W. F. (2019). Journalistic Griots: The Marginalisation of Indigenous Language News and Oral Epistemologies in Ghana. *Radio Journal: International Studies in Broadcast & Audio Media, 17*(2), 235–252. https://doi.org/10.1386/rjao_00007_1

Molale, T. B. (2019). Participatory Communication in South African Municipal Government: Matlosana Local Municipality's Integrated Development Plan (IDP) Processes. *Communicare, 38*(1), 57–75.

Molale, T. B. (2021). A Framework for Participatory Communication in the IDP Context of Ward 31 in Rustenburg Local Municipality. Unpublished doctoral thesis NWU: Potchefstroom.

Molale, T. B., & Mpofu, P. (2021). Making sense of Mmega Dikgang's Shift from Setswana to English. In A. Salawu (Ed.), *African Language Media: Development, Economics and Management* (pp. 74–89). Routledge.

Motsaathebe, G. (2010). Language, *Afrikology* and the Tremor of the Political Moment: English as the Main Language of Discourse in Africa. *Indilinga:African Journal of Indigenous Knowledge Systems, 9*(1), 96–109.

Motsaathebe, G. (2011). Journalism Education and Practice in South Africa and the Discourse of the African Renaissance. *Communication: South African Journal for Communication Theory and Research, 37*(3), 381–397. https://doi.org/10.1080/02500167.2011.605372

Motsaathebe, G. (2018). When the Subaltern Speaks: Re-examining Indigenous Language Media as an Alternative Public Sphere During Colonial South Africa. *Journal of African Media Studies., 10*(2), 169–182.

Motsaathebe, G., & Tsarwe, S. (2021). Implications of gatekeeping on talk radio. In S. Chiumbu, & G. Motsaathebe (Eds.), *Radio, Public Life and Citizen Deliberation in South Africa* (pp. 251–266). Routledge.

Mpofu, P., & Salawu, A. (2018). Culture of Sensationalism and Indigenous Language Press in Zimbabwe: Implications on Language Development. *African Identities, 16*(1), 1–16.

Mpofu, P., & Salawu, A. (2020). African Language Use in the Digital Public Sphere: Functionality of the Localised Google Webpage in Zimbabwe. *South African Journal of African Languages, 40*(1), 76–84. https://doi.org/10.1080/02572117.2020.1733833

Mugovhani, N. G. (2015). Emerging Trends from Indigenous Music and Dance Practices: A Glimpse into Contemporary Malende and Tshigombela. *Southern African Journal for Folklore Studies, 25*(Supplement 1), S81–S96.

Muturi, N., & Mwangi, S. (2009). The Theory and Practice Gap in Participatory Communication. *The Journal of International Communication, 15*(1), 74–91. https://doi.org/10.1080/13216597.2009.9674745

Nkoala, S., & Matsilele, T. (2021). In whose interest is it anyway?: An analysis of the content, discourse, and programming strategies of South African business radio shows. In S. Chiumbu, & G. Motsaathebe (Eds.), *Radio, Public Life and Citizen Deliberation in South Africa* (pp. 236–250). Routledge.

Oburu, P. (2016). *Indigenous Language Radio in Kenya and the Negotiation of Inter-Group Relations during Conflict Process.* In A. Salawu & M. B. Chibita (Eds.), *Indigenous Language Media, Language Politics and Democracy in Africa* (pp. 153–181). Palgrave Macmillan. https://doi.org/10.1057/9781137547309_8

Osunkunle, O. O. (2013). Participatory Radio As the Voice of the Community" A Comparative Study of Selected Radio Stations in the Eastern Cape, South Africa. PhD Thesis, Nelson Mandela Metropolitan University, South Africa. https://core.ac.uk/download/pdf/145053072.pdf

Oyesomi, K. O., & Salawu, A. (Eds.). (2020). *Emerging Trends in Indigenous Language Media, Communication, Gender, and Health.* IGI Global.

Reid, J., & McKinley, D. T. (2020). *Tell Our Story: Multiplying Voices in the News Media.* Wits University Press.

Salawu, A. (2019). Alaroye, Isolezwe and the Adoption of Digital Technologies. In A. Salawu (Ed.), *African Language Digital Media and Communication.* Routledge.

Salawu, A. (Ed.). (2021a). *African Language Media: Development, Economics, and Management.* Routledge.

Salawu, A. (2021b). African Language Newspaper Sustainability: Lessons to Learn from Asia. In A. Salawu (Ed.), *African Language Media: Development, Economics, and Management.* Routledge.

Tau, S. F. (2013). Citizen Participation as an Aspect of Local Governance in Municipalities: A South African Perspective. *Journal of Public Administration, 48*(1), 152–160.

The Diagram Group. (2013). *Encyclopedia of African Peoples.* New York: Routledge.

Thebe, T. P. (2016). Community Participation: Reality and Rhetoric in the Development and Implementation of the Integrated Development Plan (IDP) Within Municipalities of South Africa. *Journal of Public Administration, 51*(4), 712–723.

Tshabangu, T., & Salawu, A. (2022a). Indigenous-language Media Research in Africa: Gains, Losses, Towards a New Research Agenda. *African Journalism Studies, 43*(1), 1–16. https://doi.org/10.1080/23743670.2021.1998787

Tshabangu, T., & Salawu, A. (2022b). Technology Innovation and Digital Journalism Practice by Indigenous African-language Newspapers: The Case of *uMthunywa* in Zimbabwe. *African Journalism Studies.* https://doi.org/10.1080/23743670.2022.2049618

Tufte, T. (2017). *Communication and Social Change: A Citizen Perspective.* Polity Press.

Tyali, S. M. (2021). On Community Radio and African Interest Broadcasting: The Case of Vukani Community Radio (VCR). In W. Mano & V. C. Milton (Eds.), *Routledge Handbook of African Media and Communication Studies* (pp. 192–204). Routledge.

CHAPTER 7

# Radio and Social Media Convergence: Motsweding FM's Use of WhatsApp and the African Language Digital Public Sphere

*Ofentse Ngake, Abiodun Salawu, and Phillip Mpofu*

## Introduction: Background and Context

This chapter investigates the intersection of radio and social media, with particular emphasis on Motsweding FM's use of WhatsApp, thereby promoting indigenous African languages in the digital public realm and audience participation. Since the medium was first used in South Africa 100 years ago, the predominance of the English language on South African radio has been a matter of political, scholarly, and public concern. English has been the language that occupies more functional space in the media, broadcasting in particular at the expense of indigenous African languages (Molale & Mpofu, 2021). More so, in recent years, globalisation and information technology have allowed individuals from all over the globe

O. Ngake (✉) • A. Salawu • P. Mpofu
North-West University, Mahikeng, South Africa
e-mail: Abiodun.Salawu@nwu.ac.za



S. Nkoala, G. Motsaathebe (eds.), *100 Years of Radio in South Africa, Volume 2*, https://doi.org/10.1007/978-3-031-40706-2_7

to connect and share ideas using English, which has become the global language. For that reason, African languages play a peripheral role in the African public domain, especially in the digital public sphere. This has significantly affected the participation of ordinary people in the online public sphere, notably those who are not conversant with the English language. Thus, language has become a barrier for people's participation in the digital public sphere. The global technological wave supported by the Internet and social media has profoundly transformed South African communication systems. This raises concern about the survival and maintenance of indigenous African languages in a globalised and digitalised world, as well as ordinary people's participation in the digital public sphere.

However, in a democratic society media participation is key since it allows public participation in radio and television programmes that may be of local, regional, or national interest. Listeners to the radio, in particular, find it exciting and convenient to engage in and respond to the numerous themes offered on various programmes. Since language is a primary means of communication, using indigenous African languages enhances communication and participation in radio conversations.

However, social media has become a crucial component of radio stations' operations and functions (Gavaza & Pearse, 2019; Matsilele et al., 2023). The convergence of traditional media with new media technologies has seen radio extending its presence on online platforms, including social media (Molale & Mpofu, 2023). Radio has converged with social media, thereby remaking the functioning and operations of radio, especially in terms of audience engagement and participation. Social media platforms have made it easier for audiences to participate in the digital public realm. This has facilitated the audience to engage with radio and contribute to radio programmes using indigenous African languages. Deploying different social media platforms, topics are posted to allow audience reactions through liking, commenting, or sharing. These reactions can then be relayed live on air by radio presenters or read by netizens.

The public sphere is a space where citizens connect and communicate issues of public concern (Gripsrud & Moe, 2010). Before the advent of the digital age, audience participation was limited. Radio was one of the earliest platforms used by citizens from all walks of life to communicate and participate in the public sphere. For their involvement in radio shows, people used to write letters, which made it relatively slow, and often these letters were not read. According to the African National Congress (2010, p. 7), media participation pre-1994 was characterised by the inequality of access, language, and cultural and educational programming. Digital media technologies are now integrated into much of what people do and have improved interaction and communication (Johann et al., 2016). They have improved people's access to information at any time and in any place. With social media and Web 2.0 having emerged, the digital public sphere has changed into a more open communication space. People can now use social media, emails, and online streaming to participate in radio programmes. Audience participation has developed gradually since the existence of the digital world.

According to Galal, South Africa has 38 million Internet users, with WhatsApp accounting for 95% of the online audience. While the Internet is said to be dominated by the English language (Nkoala, 2020), social media are also viewed as breathing spaces for minoritised languages (Belma & Glass, 2019). Also, Galla (2016) maintains that it is quite challenging to imagine the survival of indigenous languages in the twenty-first century without the intervention of digital technology in a culturally diverse world. The rationale is that indigenous people can use digital tools to supplement language documentation, revitalisation, promotion, and education efforts.

Against this background, this chapter appraises the use of social media, particularly the WhatsApp voice note facility, by Motsweding FM, in engaging with listeners and how that interface promotes the use of indigenous African languages on radio and the African language digital public sphere. According to the WhatsApp website, the application sends and receives various media: text, photos, videos, documents, and location, as well as voice and video calls. In addition, this application has a feature of sending and receiving voice recordings called voice notes. The user records their voice message and sends it to the recipient instead of typing it.

Motsweding FM is a South African national radio station based in Mahikeng, North West Province, that transmits in Setswana. It transmits throughout the North West, Northern Cape, Gauteng, and Free State, and is expanding into Limpopo and Mpumalanga. It also streams online on its website and is available on the DSTV audio bouquet, Channel 408. Setswana is one of the eleven official South African languages widely spoken in the North West province. The station is one of fifteen that are South African Broadcasting Corporation (SABC) owned. It has been broadcasting since 1962 when it was Radio Tswana operating from Mmabatho in the former Bophuthatswana homeland. The station's tagline is *Konka Bokamoso*, which can be loosely translated into Realising the Future. The station describes itself as "the contemporary voice of Setswana speaking and understanding South Africans, acknowledging, proud and respectful of its cultural heritage". As one of SABC's Public Broadcasting Service Radio stations, its line-up features include diverse programmes, news, current affairs, and music. Of interest in this chapter is a programme called *Resemeletse*, which literally means "we are working very hard". Focusing on the voice note service, this chapter explores how this WhatsApp application promotes the use of indigenous African languages and participation in the digital public sphere.

The chapter begins by situating use of WhatsApp as an application used by radio stations, highlighting the scarcity of scholarship exploring this phenomenon in the South African context. It proceeds to explore literature on social media and the promotion of indigenous African languages in the public sphere. This is done by reviewing existing studies on the radio medium, social media, and the need to promote indigenous African languages in the media. The chapter then elaborates on the theoretical framework deployed in the study, before outlining and discussing the key findings of the study. It concludes by demonstrating the central argument advanced and the contribution of the study to the existing body of scholarship on media convergence, radio broadcasting, and social media in particular.

## Literature Review

Existent studies show that radio remains the primary mass medium, with the most extensive geographical reach and largest listener numbers since it facilitates the flow of information across communities and is an established instrument of communication. Its place in South African society has remained firm over the 100 years of its existence. Still, it has not been unaffected by emerging media, such as social media, which have changed the media ecosystem.

Despite the increased popularity of visual and social media, radio continues to play a significant role among South Africans. This could be explained by radio's distinct advantage over other forms of media, such as the appeal of radio music and soap operas, among other things. While new technology is thought to replace whatever medium, radio has remained in use (Danos & Turin, 2021). It is a valuable medium for transmitting news, songs, and debates to communities. Radio is also the principal source of entertainment for many people in South Africa. African language radio, as a component of African language media, should be more efficient in strengthening the media's entertaining, instructive, instructional, and iconic purposes. Radio is also an efficient and inventive tool for the revitalisation and reclamation of indigenous languages.

Existing scholarship shows that radio broadcasting is increasingly converging with social media (Molale & Mpofu, 2023). In a study on the relationship between social media and radio, Karttunen (2017) shows that social media play a pivotal role as a publishing platform for radio stations and a tool for audience participation. Karttunen (2017) further explores advantages and disadvantages of using social media by radio stations and advocates for radio stations to have social media platforms. Radio stations should have social media and social media strategies at hand as a guideline for the success of the station. While Karttunen (2017) considers social media broadly, and implores radio stations to have social media accounts, this chapter focuses specifically on the use of WhatsApp voice notes by an African language radio—Motsweding FM. In the context of English's dominance in the online space, it is critical to examine the use of WhatsApp in promoting indigenous languages using African language radio in the digital public sphere (Mpofu & Salawu, 2020).

Sherwood (2015) explores how radio stations adapt and are going beyond the traditional way of broadcasting and letting their listeners participate on radio, that is, how radio has evolved to improve audience participation in radio programmes. Radio stations have introduced social media platforms as ideal means for audience participation. Radio stations have mostly edged into social media to connect with listeners and reach beyond the dial. The radio and social media convergence aptly shows that radio stations are adopting social media as a medium of audience participation. This view demonstrates social media's expediency in enhancing audience participation. Radio stations in South Africa, including Motsweding FM, have adopted the use of social media, as demonstrated by the existence of Facebook and Twitter accounts. Motsweding FM uses the WhatsApp application to facilitate audience participation in on-air engagements. This chapter appraises this development and analyses how this promotes indigenous African languages and the participation of the audience in the digital public sphere.

The promotion of indigenous languages in the public domain, including the indigenous language digital public sphere, has been highlighted and underscored in existing literature (see Mpofu & Salawu, 2020; Nkoala, 2022). Benson et al. (2017) explore the issues and challenges associated with promoting indigenous languages by library and information sciences professionals. The mentioned scholarly work further argues that promoting indigenous languages should start from the grassroots by encouraging the writing, translation, and teaching of indigenous languages. Kaya et al. (2016) looked at the role of indigenous African languages in higher education transformation, with a specific slant on the challenges higher education lecturers face in using indigenous languages in digital learning environments. To sustain indigenous languages, a mass of young, committed experts should be built to promote indigenous languages. This demonstrates the imperative for promoting indigenous African languages in the public domain.

In particular, Mpofu and Salawu (2018), illustrate the position and utility of African language media, emphasising the importance of giving African languages broadcasting space. Use of indigenous African languages enhances the efficiency with which media systems convey information while increasing audience participation. Also, African language media maintain regional and national identities. In the South African context, broadening the space for African languages in the media decolonises African media systems which are traditionally dominated by ex-colonial languages such as English, and the language of the apartheid regime, Afrikaans. This means that using African language media helps to promote the execution of the media's entertainment, instructive, and educational aims.

## Theoretical Framework

To appreciate the use of social media, in this case WhatsApp by Motsweding FM, to interact with listeners and how that interface promotes an African language digital public sphere, this chapter is grounded in the theoretical concepts of convergence, diffusion of innovation, and the digital public sphere. Convergence entails the flow of content across multiple media platforms, which is the cooperation between multiple media industries (Jenkins, 2006). This is the process by which new technologies are accepted by current media and communication sectors and cultures (Dwyer, 2010). It is the incorporation or integration of mass communication media such as radio, television, the Internet, and newspapers, as well as portable technologies and other interactive devices via various digital presentation platforms or other languages, thereby combining different media into a single medium (Gani et al., 2019).

Media convergence comes from the growth of the Internet, which allows consumers to access everything easily and everywhere. The term convergence refers to the adaptation and fusion of old and new technology. Convergence modifies the interactions between technologies and audiences (Jenkins, 2006). In this study, the deployment of WhatsApp by Motsweding FM demonstrates the convergence of broadcast media and social media. Convergence allows an understanding of how the digital world is incorporated into traditional broadcast media usage norms for audience participation. In this study, convergence provides insights into the role played by social media in promoting indigenous language digital media. Over the past few years, the radio industry has broadcast in the traditional way. Since the existence of the digital world, media platforms such as radio that were not digital have now had to incorporate digital media platforms. Considering the large number of online listeners, the adoption of social media by radio stations is bound to widen listenership. It has also improved the process of audience participation, allowing listeners to post content, and for their posts to be read online if not read on radio.

Diffusion is a process through which innovation is communicated and channelled over a social system (Rogers, 2010). Diffusion happens when there is innovation to diffuse. In terms of innovation, it is referred to as new ideas or methods that are used. In the context of this study, the WhatsApp service is the innovation. It is an innovation in the sense that radio has been broadcasting the conventional way before the new innovation: WhatsApp. WhatsApp was adopted to interact with the audience and promote an indigenous African language digital public sphere. Adopting an innovation and its usage is affected by factors such as relative advantage, compatibility, and complexity (Jeong & Lambert, 2001). These elements contribute to the innovation's effect on being thought of as superior to the concept or procedure it replaces. The usability of an innovation also counts as this determines the innovation's adaptability. As a result of its relative advantage, compatibility, and adaptability, social media, including WhatsApp, have been embraced by broadcast media.

The digital public sphere is a communication platform achieved through social media and other online communicative platforms, where matters of mutual concern are deliberated on and made visible to all (Schäfer, 2015). In this study, the digital public sphere refers to the online space where the radio station Motsweding FM interacts with the audience using the WhatsApp facility. Deploying this concept, this study explores the contribution of WhatsApp in promoting an indigenous African language digital public sphere, thereby promoting audience participation.

## Methodology

This study adopted a qualitative approach which was interpretive in nature. This allowed researchers to interact with the participants and understand the investigated social phenomena. Specifically, this is a case study of Motsweding FM's *Resemeletse* programme's use of WhatsApp. The study population consisted of thirty-two participants comprising thirty listeners of the programme, as well as the presenter and the producer of the programme. The study population was chosen because of their frequent participation in the examined radio broadcast, with the producer and presenter participating in production and the audience in reception. The participants were selected using the purposive sampling technique for all population segments. This sampling technique allowed for the selection of listeners according to their use of WhatsApp as a form of participation on radio, while the producer and the presenter were selected because of their involvement in the production of Motsweding FM's *Resemeletse* programme.

The study deployed three primary data collection methods: interviews, questionnaires, and observation. Data was collected from the presenter of the *Resemeletse* show and the producer of Motsweding FM using unstructured interviews. Data from the listeners was collected using both interviews and questionnaires. Data was also collected through non-participant observation by following the interaction between the radio presenter and the audience. Observation is a purposeful way of watching and listening to an exchange as it takes place. Researchers also observed how the presenter administered the programme using WhatsApp voice notes on air to engage and communicate with the audience. The study was carried out following all necessary ethical considerations.

## Findings and Discussion

Research findings show that among the listeners who interact with Motsweding FM on the programme *Resemeletse*, the majority use the WhatsApp messaging application. The use of WhatsApp is an innovation used by the radio station, and most of the listeners who participated in this study used it to participate in the studied programme. Furthermore, interviews with the presenter and producer of the programme reveal that they considered the utility and accessibility of the WhatsApp messaging service to connect and interact with the audience. Similarly, audiences' responses also show that they consider the application's convenience for interacting with the presenter on the radio station.

The aforementioned finding demonstrates that social media has had an impact on radio transmission. That is, despite the rising popularity of social media, radio has continued to play an important role in South African communities. Instead, radio and social media have merged to create an intriguing digital public space. This finding concurs with Danos and Turin's (2021) observation that while new technology is generally thought to replace traditional media, radio has remained in use. In fact, radio broadcasting is increasingly converging with social media since radio stations have adopted and adapted social media to their operations (Sherwood, 2015) (Matsilele et al., 2023). It is also evidence of the adoption and adaptation of a new innovation (WhatsApp), which injects a fresh creation into the operations of radio stations, especially the engagement and participation of the audience. This reinforces the assumption that social media is becoming an important instrument for audience interaction (Karttunen, 2017). In reality, the convergence of technology encourages audience engagement.

While text messages were used by the audience, the survey findings suggest that WhatsApp voice messaging was the most popular among Motsweding FM listeners. WhatsApp voice messaging allows users to interact instantaneously by sending recorded messages. As a result, audio messages are recorded by radio listeners and sent to the radio station's WhatsApp account. The host then plays the messages on air, reaching the audience in a manner similar to phone-in programmes. The thrill and satisfaction of hearing their voices on the radio, mostly in Setswana, is enjoyed by the listeners. The participants' perceptions show they are thrilled to use WhatsApp voice messaging to participate on the radio. They share the same sentiments with the presenter and producer on the affordability of the service as compared to the telephone. This method's immediacy is comparable to traditional telephone interactions. This supports Jenkins' (2006) claim that convergence alters the interactions of technology and audiences.

The study also discovered that Motsweding FM, an African language radio station, uses WhatsApp for text and voice messaging, which encourages the usage of Setswana in presenter and listener interactions, hence developing an indigenous African language digital public sphere. Each day, *Resemeletse* programme listeners contribute to the radio station using WhatsApp text and voice messages. The producer and presenters at Motsweding FM deem the app to be extremely beneficial and effective for both them and the audience at large. Participants are regular individuals, and their voices are heard. This shows that WhatsApp is a digital public space where radio, an indigenous African language media, has expanded its presence. As a result, even if English is believed to dominate online spaces (Mpofu & Salawu, 2020; Nkoala, 2022), the usage of WhatsApp supports African language radio, thus increasing the indigenous language's digital public domain. This supports Belmar and Glass' (2019) claim that social media platforms provide breathing room for minority languages. While scholarship has focused on increasing the African language broadcasting space (Mpofu & Salawu, 2018), this chapter has demonstrated how social media may be used to promote indigenous language radio in the digital public realm.

## Conclusion

With a special emphasis on Motsweding FM's usage of WhatsApp, this chapter offers insights into the dynamics and affordances of South African radio and social media convergence. Traditional radio has expanded its online presence as a result of the rise of social media. The radio industry has changed as a result of this convergence, notably in terms of listener engagement. Social media platforms have made it easier for audiences to participate in the digital public realm. This chapter has shed light on the long-term viability of radio transmission in the context of social media. Against the backdrop of South Africans' rising interest in and usage of social media, as well as the convergence of radio and social media, radio stations' adaptation to the use of social messaging apps such as WhatsApp demonstrates the medium's durability in the digital era. This chapter demonstrates how the convergence of radio and social media, in general, and the use of WhatsApp text and voice messaging functions, in particular, promotes the expansion of the indigenous African language public sphere against the backdrop of the global technological wave supported by the Internet and the supremacy of the English language in South African media and the digital space.

## References

African National Congress. (2010). Media Transformation, Ownership and Diversity (p. 7).

Belmar, G., & Glass, G. (2019). Virtual Communities as Breathing Spaces for Minority Languages: Re-framing Minority Language Use in Social Media. *Adeptus*, 1–24.

Benson, O. V., Anyalebechi, L. I., & Ariole, I. A. (2017). Promoting Indigenous Language. In Nigeria: Issues and Challenges for the Library and Information Professionals. *Library Philosophy and Practice*, 1–15. Retrieved January 30, 2022, from http://digitalcommons.unl.edu/cgi/viewcontent.cgi?article=4162&context=libphilprac

Danos, D., & Turin, M. (2021). Living Language, Resurgent Radio: A Survey of Indigenous Language Broadcasting Initiatives. *Language Documentation and Conservation, 15*, 75–152.

Dwyer, T. (2010). *Media Convergence*. Open University Press.

Galla, C. K. (2016). Indigenous Language Revitalisation, Promotion, and Education: Function of Digital Technology. *Computer Assisted Language Learning, 29*(7), 1137–1151.

Gani, N. A., Maulianza, M., & Mulya, M. R. A. (2019). *Social Media and Radio Broadcasting Integration to Expand the Market of IRadio Jakarta* ICEMA International Conference on Economics, Management and Accounting Volume 2019.

Gavaza, M. A., & Pearse, N. J. (2019). Assessing the Social Media Maturity of a Community Radio Station: The Case of Rhodes Music Radio in South Africa. *The African Journal of Information and Communication (AJIC), 24*, 1–21. https://doi.org/10.23962/10539/28661

Gripsrud, J., & Moe, H. (2010). *The Digital Public Sphere: Challenges for Media Policy*. Coronet Books Incorporated.

Jenkins, H. (2006). *Convergence Culture: Where Old and New Media Collide*. New York University Press.

Jeong, M., & Lambert, C. U. (2001). Adaptation of an Information Quality Framework to Measure Customers' Behavioural Intentions to Use Lodging Web Sites. *International Journal of Hospitality Management, 20*(2), 129–146.

Johann, D., Greff, T., & Werth, D. (2016). On the Effect of Digital Front Stores on Transforming Business Models—Concept and Use-Case from the Consulting Industry. Retrieved January 31, 2022, from https://www.researchgate.net/profile/Dirk_Werth/publication/307608417_On_the_Effect_of_Digital_Frontstores_on_Transforming_Business_Models_-_Concept_and_Use-case_from_the_Consulting_Industry/links/57cd6c6e08ae5982518a1d5f.pdf

Karttunen, S. (2017). Using Social Media at a Radio Station. Unpublished undergraduate research paper. University of Applied Sciences, Helsinki Metropolia.

Kaya, H. O., Kamwendo, G. H., & Rushubirwa, L. (2016). African Indigenous Languages in Higher Education. *Studies of Tribes and Tribals, 14*(2), 121–128.

Matsilele, T., Makwambeni, B., Nkoala, S., & Bulani, G. J. (2023). Youth Audiences and Social Media Integration in Community Radio Stations in South Africa: A Case Study of Zibonele FM and Bush Radio. In S. Tsarwe & S. Chiumbu (Eds.), *Converged Radio Youth and Urbanity in Africa. Emerging Trends and Perspectives* (pp. 123–153). Springer International Publishing.

Molale, B. T., & Mpofu, P. (2021). Making Sense of the South African Mmega Dikgang's Transition from Setswana to English. In A. Salawu (Ed.), *African Language Media: Development, Economics and Management*. Routledge.

Molale, B. T., & Mpofu, P. (2023). (Dis)continuities of African Language Radio on Social Media: The Case of South Africa's Motsweding FM and Radio Zimbabwe. In P. Mpofu, I. A. Fadipe, & T. Tshabangu (Eds.), *African Language Media*. Routledge.

Mpofu, P., & Salawu, A. (2018). Culture of Sensationalism and Indigenous Language Press in Zimbabwe: Implications on Language Development. *African Identities, 16*(1), 1–16.

Mpofu, P., & Salawu, A. (2020). African Language Use in the Digital Public Sphere: Functionality of the Localised Google Webpage in Zimbabwe. *South African Journal of African Languages, 40*(1), 76–84.

Nkoala, S. B. (2020). Student perceptions of multilingualism and the culture of communication in journalism studies in higher education. *Reading & Writing, 11*(1), a258. https://doi.org/10.4102/rw.v11i1.258

Nkoala, S. (2022). Educators' Experiences of Using Multilingual Pedagogies During Emergency Remote Teaching: A Case Study of South African Universities. *International Journal of Multilingualism*, 1–14. https://doi.org/10.1080/14790718.2022.2074012

Rogers, E. M. (2010). *Diffusion of Innovations* (4th ed.). The Free Press.

Schäfer, M. S. (2015). Digital Public Sphere. In *The International Encyclopedia of Political Communication* (pp. 322). Wiley-Blackwell.

Sherwood, A. (2015). (Re)Connecting with Listeners: How Radio Stations are Reaching beyond the Dial (and Their Competitors) to Connect with Their Audience. Unpublished master's dissertation. RIT Main Campus, Rochester Institute of Technology.

CHAPTER 8

# Radio Production in the Digital Era: Lessons From South Africa

*Carolyne M. Lunga* and *Rosemary K. Musvipwa*

## Introduction

Digital technologies have altered how media organisations operate globally. The advent of digitalisation has resulted in increased adoption of online platforms by media entities seeking to enrich their stories and promote interactivity with their target audiences. Radio stations in South Africa are no exception, as media practitioners in various entities increasingly use digital media to practice their craft and reach new audiences (Bosch, 2010, 2014). Audiences use these platforms to engage with the radio stations and participate in critical issues of the day (Wake & Bahfen, 2016)(Matsilele et al., 2023). Access to smartphones and internet penetration has resulted in more people tuning into digital formats (Nkoala,

C. M. Lunga (✉)
University of Portsmouth, Portsmouth, UK
e-mail: Carolyne.Lunga@port.ac.uk

R. K. Musvipwa
University of Eswatini, Kwaluseni, Eswatini
e-mail: rmusvipwa@uniswa.sz



S. Nkoala, G. Motsaathebe (eds.), *100 Years of Radio in South Africa, Volume 2*, https://doi.org/10.1007/978-3-031-40706-2_8

2023; Bosch, 2010). Radio has, over the years, continued to be the most pervasive electronic communications media, leading to the development of new technologies to support the demand for radio services (Coyne, 2010; Verdone & Zanella, 2012). This change has been accompanied by a renewed interest in research on radio in the digital era, often conceptualised in terms of how it promotes the public sphere and provides a space for previously marginalised groups to express themselves. These studies are often linked with the notion of democracy, citizenship and identity rooted within Western scholarship while analysing local contexts and realities. We note that it is important for research done in the Global South context to consider how locals exercise democracy, citizenship and identity and interrogate the heterogeneous ways in which these are shaped by the political, economic, social and media contexts. This is important to provide insights, knowledge and understanding of digital media practices outside the Global North. Research on countries in the Global South has often been marginalised and lacking in literature (Mutsvairo, 2019).

There is evidence of research in both the Global North and South to show how media production, dissemination and consumption processes are to be understood in this context of digital technologies in relation to local contexts (see Ndlovu & Sibanda, 2021; Ndzinisa et al., 2021; García-Avilés, 2021; Verweij, 2009; Chiumbu & Ligaga, 2013). Existing research on radio has considered the role of radio as a public sphere or a space for counter-discourses and analysing issues of access and participation and the early years of broadcasting (Chiumbu & Ligaga, 2013; Bosch, 2014; Mhlambi, 2008). Existing research has interrogated how radio developed during colonial times and how it was used as a form of control (Mhlambi, 2008), how Radio Freedom was used to challenge apartheid (Tyali, 2020), the development of community radio (Bosch, 2010), the role of community radio within rural communities (Fombad & Jiyane, 2019), and the ability of South African talk radio stations to fulfil their normative social responsibility roles (Nkoala & Matsilele, 2021) among others.

These studies show the extensive research in existence when it comes to radio studies in South Africa. In their research, Chiumbu and Ligaga (2013) analysed how new media technologies transformed radio practices in South Africa in terms of institutional cultures and audience engagement. Their study drew on the public sphere theory and demonstrated that radio created a public that produced, circulated and shared information in various ways. The study also acknowledged the limitations of access within this context because of the digital divide. Within the African context, questions regarding power, access and inequality are still being asked. This is because even where new digital technologies are being appropriated by various media and audiences, they are still dominated by the elite, who can create new online communities and participate online, thereby excluding those without access and lacking the media literacy skills to participate on these platforms (ibid.).

This chapter analyses similar issues regarding audience engagement and new production and dissemination practices but differs in approach in several ways. Firstly, we draw on the social constructivism theory and use four different radio stations to analyse how they are appropriating technologies in newsgathering, production, dissemination and processes of engaging with their target audiences. The four stations are all South African based. Radio 947, Kaya FM and 5FM are Johannesburg based stations while Ukhozi FM broadcasts from KwaZulu Natal. All stations have adopted livestreaming capabilities and have developed a heavy online presence constituted of various platforms, tools and techniques. As we conducted the research, we acknowledged that the Covid-19 pandemic could bring about some changes. This is important to highlight as we have witnessed some media globally and in South Africa close after suffering from massive economic challenges due to the breakdown of the business model amid the pandemic. On the same note, the Covid-19 era has seen more media organisations intensify and strengthen their digital presence to cater for physically dispersed audiences. In the South African context, radio broadcasts were used as an educational tool for young people with no access to the internet (Bosch, 2010). We thus ask how the four South African radio stations, namely radio 947, Kaya FM, UKhozi FM and 5FM, operate within this context to shed light on how they are transforming.

We begin by providing an overview on the development of radio in South Africa, in light of its 100th year of existence in this context. We then consider the four stations of interest and give context to why each of these is relevant when considering radio in South Africa in the digital age. The chapter then discusses why social constructivism theory is an appropriate lens through which to consider the issue of radio production in South Africa in the digital age. Our discussion section highlights findings in relation to the strong online communities that have been created as a result of the use of digital technologies by these stations. It then discusses the interplay between culture and community in these online contexts. It expounds on the role of audiences as co-creators of content considering the affordances presented by digital tools, and finally talks about radio and the digital public sphere. In doing this, our overarching goal is to demonstrate that the four stations are appropriating digital technologies in relation to their contexts.

## The History and Development of Radio in South Africa

Radio is considered "Africa's medium" and is often described as the most powerful tool on the continent. It is considered the most widespread and affordable medium accessible to most populations in rural and urban areas, both literate and illiterate. Nkoala (2023) argues that in the South African context, radio remains the most loved medium, and its intimacy, immediacy and sociability are unmatched. Its ability to transcend borders also contributes to its popularity (Evans, 2022). As in most African nations, the history of radio in South Africa is tied to its colonial legacy, as it was used as a propaganda tool to control what native audiences were exposed to (Mhlambi, 2008; Nkosi, 2014; Tyali, 2020). In addition, it was used to instruct black South Africans to conform to the policies of the apartheid government and to portray themselves locally and internationally (Nkosi, 2014; Tyali, 2020). Bosch notes that the history of radio in South Africa is closely intertwined with the country's struggle, political transition and dynamics of contemporary society. Radio is argued to have begun in South Africa in 1923 and was taken over by the State in 1936 when it became the South African Broadcasting Corporation (SABC) (Gunner, 2000). It is argued to have created "cultures of listening and free sites of the imagination, in spite of an oppressive state radio policy", as considered of the SABC in the apartheid era (Gunner, 2000, 2017). Unlike newspapers, which require some degree of literacy, radio could be easily accessed by anyone who could listen; thus, it could effectively capture a vast base of listeners (ibid.).

Radio in South Africa has gone through various stages. In April of 1941, journalist King Edward Masinga made a one-minute Zulu-language report of war news from the SABC studios in Durban. It is argued that "the space was an ideologically charged one and it clearly marked the culture of radio, particularly African-language radio, as a site of struggle" (Gunner, 2000, pp. 223–224). Radio freedom is argued to have played a critical role in advancing the anti-apartheid struggle (Tyali, 2020). Bosch (2006, p. 250) notes that:

> Except for the ANC's Radio Freedom broadcasting from exile, the democratic movement ignored broadcasting as a site of struggle. Radio Freedom was the underground radio station of the ANC. Founded in 1967 [?]It broadcast into South Africa daily on a shortwave from neighbouring African states, particularly Zambia and Mozambique. As part of the ANC's cultural wing, Amandla, Radio Freedom provided the only alternative to the strongly censored SABC, merging political content and news with popular music of many banned artists.

The above quote demonstrates the important role played by Radio Freedom in clamouring for democracy. The airwaves in South Africa were opened for the first time in history when the Independent Broadcasting Authority was created on 30 March 1994 by an act of parliament (Mtimde, 2003). This marked the beginning of the democratic dispensation (Mtimde, 2003). When apartheid came to an end in the 1990s, South Africa adopted a three-tier broadcasting system made up of public, commercial and broadcast media (Mtimde, 2003). Since then, the broadcasting industry has continued to grow to cater for the population needs of South Africa. In the present context, the radio industry has seen the growth of partnerships and the creation of solutions which offer partners and audiences benefits in an increasingly competitive environment. Collaborations, offering diverse content, live streaming and mixing of platforms, feature top on the agenda of most stations.

## Radio in South Africa in the Digital Era

The digital era of radio in South Africa is characterised by various shifts in institutional and audience engagement practices. Some scholars perceive new technologies as a threat to the functioning of the media, while others consider them positively in enhancing the work of journalists and media practitioners and allowing them to build stronger relationships with their audiences while enhancing democracy. Radio plays an important role in enhancing democratic culture and space. In this chapter, we argue that digital platforms strengthen the station's offerings, which bring about co-creation/co-production with audiences and create overlapping local, national and global communities online. The fluidity of radio and its connection with new digital platforms enables it to extend its reach and connect remote and urban-based communities (Gunner et al., 2011).

Mobile technology, for instance, has extended the reach of radio, enabling audiences to interact with the stations instantaneously (Chiumbu

& Ligaga, 2013; Tettey et al., 2011). As a result, audiences take up various roles and become part of the "imagined centre" of the time (Gunner, 2017, p. 62). Moreover, the use of multiple "knowledge enabling formats make complex politics more intelligible to a wider audience" (Davis, 2010). Tettey et al. (2011) argues that participation in radio demonstrates agency and is important for citizenship and democracy, seen through talk radio's promotion of inclusive discourse with audiences who can combine vernacular and English and set the agenda by calling power to account. Radio is also commended for bringing previously marginalised voices into the public sphere (Gunner et al., 2011; Tettey et al., 2011).

Amid this, it must be noted that the adoption of digital technologies for radio production purposes has not been uniform. As with the staggered introduction of radio to different audiences over the past 100 years, issues of access and affordability have meant that some stations and their listeners have been able to incorporate digital technologies more readily than others. Hence the approach employed in this study has been to consider four different cases and the South African media context. For example, "information access inequalities and new media literacy compromise the representativeness of the virtual sphere" (Papacharissi, 2002, p. 9). Not all citizens have access to online technologies and have the necessary English language skills required to interact on these platforms whose content is mainly in English.

The first station, Ukhozi FM, is owned by the SABC and is one of the oldest radio stations in the country. It is the only indigenous language public service broadcasting station considered in the study. The second station, 5FM, is also owned by the SABC and broadcasts in English and is one of three SABC commercial stations, along with Metro FM and Good Hope FM. Radio 947 is a regional commercial station broadcasting in Gauteng and it is owned by the Primedia Group. Kaya FM is also based in Gauteng and is owned by Kagiso Media. Issues of ownership and management are central when determining the extent to which digital technologies feature as part of a particular radio station's production practices. While listeners and presenters may organically adopt these tools, for them to be used at the more strategic levels of how the station operates, those tasked with leading radio stations often have an important say. This issue, however, is not the focus of this particular chapter and should perhaps be an area of future study for researchers.

### *Ukhozi FM*

Ukhozi FM is an SABC station which has been in existence for 62 years. The station currently has a presence on various online platforms, including Twitter, Facebook and YouTube, and its live stream is easily accessible on a laptop/desktop or mobile phone with internet connectivity. This enables broadcasting nationally and live streaming to the world (Ukhozi FM website). The station defines itself as "one of the biggest radio stations on the planet and the largest in Africa, with its listenership in constant excess of 7.7 million over the past decade" (UKhozi FM website). The station broadcasts mainly in IsiZulu and loosely targets IsiZulu-speaking and -understanding audiences in South Africa (ibid.). Gunner (2017, p. 63) argues that:

> isiZulu which presenters and callers-in use ranges from formal and even slightly old-fashioned to hip, colloquial and laced with neologisms and at times regional accents and idioms. For many listeners their connection to Ukhozi makes them part of a moving project, cogs in a process of trying to build a community, hold a life together, dream in a time of hardship.

Ukhozi FM provides an interactive environment for its listeners, affording them access to news, current affairs, talk shows, music, drama, sport, education, weather and traffic, with much emphasis on local content. Ukhozi FM has radically changed in the "new dispensation" of post-1994 South Africa, yet has maintained the creative and resistant identity forged in the apartheid years during its era as Radio Zulu (Gunner, 2017).

Ukhozi FM's headquarters is in KwaZulu Natal, Durban, and the station caters to people ranging from young to elderly, specifically the youth, reinforcing a sense of pride and culture in the young people of South Africa (Ukhozi FM website). The station keeps its listeners connected to their cultural identity in a modern world context and focuses on edutainment and infotainment as a guiding philosophy (ibid.). Gunner (2017, pp. 62–63) argues that it embraces its past as an "apartheid-era radio station through celebrating its famous—and in the case of Masinga, controversial—early broadcast figures such as King Edward Masinga, Winnie Mahlangu, Hubert Sishi, Thokozani Nene and others. It creates its own genealogy and seeks to carry the creative—and resistant—radio achievements of such personalities and of the station itself into the post-1990 broadcasting era".

### *5FM*

Like Ukhozi FM, 5FM is an SABC-owned award-winning station that broadcasts nationwide and streams worldwide (5FM website). The station calls itself the "entertainment powerhouse for South African youth, offering the most popular contemporary music and entertainment and known for its music, DJs and events" (5FM website). Headquartered at the South African Broadcasting Corporation (SABC) in Auckland Park, Johannesburg, the station "offers audiences access and opportunity to be part of the energy and dynamism of the fast-changing youthful global community" (5FM website). The station runs shows and competitions and offers listeners the opportunity to submit their favourite music, which is one of the ways of engaging citizens in production processes. The station has a presence on Facebook, Twitter, Instagram, YouTube and TikTok, which enables it to cater to the needs of diverse groups of young people in South Africa. It has close to one million weekly listeners (BRC RAM, 2021), more than 300,000 likes on Facebook, and close to one million followers on Twitter, and close to two million people follow 5FM DJs on Twitter. Listeners are able to engage with the platform via email and SMS communication. The "trending on 5FM" section on its website allows users to see shows/programmes that are popular on the platform.

### *Radio 947*

Radio 947 is a contemporary radio station that broadcasts on the 94.7 FM frequency in Johannesburg. It is based in Johannesburg in the Gauteng Province of South Africa and is privately owned by Primedia (947 website). The station has been running since 1964 and started as Radio Highveld Stereo, the first SABC regional FM service, until it was sold to a private entity in 1996. In 2014 the radio station dropped the name Highveld Stereo and became known as 947 (947 website, 2022). Radio 947 has a website and is present on Facebook, Twitter, Instagram and YouTube. The website showcases 947 Top 40 music, latest news and shows accompanied by images of the presenters behind them. Some of these presenters include sports personality Robert Marawa and media personality Anele Mdoda. Additionally, the station broadcasts via a live stream on its website and a mobile application. Operating under the tagline "947 loves you", the station broadcasts in English and provides an urban mix of programming, which includes news, music, sports, competitions, informal discussion/talk shows and traffic updates (947 website, 2022). Radio 947 has approximately 1,055,000 listeners per week, making it the biggest regional station in Gauteng (BRC RAM, 2021).

### *Kaya FM*

Established in 1997, Kaya FM is a commercial radio station like 5FM and 947. Thebe Investment Corporation has the major shareholding in the station (Thebe.co.za, 2022). It broadcasts from Johannesburg in the Gauteng province of South Africa and has a growing audience of 25 years and older, which they term "Afropolitans". With approximately 777,000 listeners per week, Kaya FM airs on the 95.9 FM frequency (Kaya FM website, 2022). The station connects with its "loyal listeners through a range of digital and on-air platforms" (Kaya FM website, 2022), including a live stream, Facebook, Twitter, YouTube and Instagram, and a mobile application. The station's programming includes news, music, sports and current affairs (Kaya FM website, 2022). Its website is dominated by news and entertainment which are shareable to email or various social media platforms such as WhatsApp, Twitter, Facebook, and LinkedIn. The station broadcasts in English and isiZulu. To enhance the listeners' experience, the station combines soul and R&B, kwaito, jazz, chart-toppers and old school jams, and boasts of presenters who are big names in the media industry. We now move on to social constructivism theory, which is the theory that informs our study.

## Social Constructivism

This chapter is informed by social constructivism theory. Social constructivism allows for the analysis of how the stations under study are appropriating digital tools and social media platforms within their contexts (Ndzinisa et al., 2021; Ndlovu & Sibanda, 2021). As argued by How (2010, p. 1), "technology and society are inseparable", and technology often gets entertwined with various social issues such as gender, race, ethnicity and religion. Social constructivism considers the relationship between technology and social life (O'Gorman, 2018). New technologies are "harnessed and domesticated" by media professionals within "their local circumstances" (Mabweazara, 2015, p. 13), which is vital for us in exploring the way they are being used by the different stations. By analysing the local context we are able to explore the situated realities of new technologies (Bosch, 2010) and provide a more nuanced understanding without generalising or uncritically using Western perspectives on the usage of these technologies. We move away from using technological determinism, to avoid an approach which looks at "what media technologies cause to how media technologies function within social life" (O'Gorman, 2018, p. 2). In addition, the technological determinist approach has often been criticised for being too technology centred and thus ignoring the lived experiences of those using it. Social constructivism is a useful approach for a study like this one which focuses on the tools and platforms and how they are being appropriated by the four stations in relation to their unique contexts. In the South African context, the digital divide has led to access challenges, particularly for those from impoverished areas of the country, while those who are well-off have regular access to digital technologies.

## Methodology

To answer the main research question of how selected South African radio stations navigated convergence and digitisation within the radio industry, the study is located within the qualitative approach and adopts a comparative case study and radio content analysis of four stations: Radio 947, Kaya FM, Ukhozi FM and 5FM. The four radio stations were chosen because they are listed as being among the top 20 radio stations in the country (Briefly, 2021). The comparative case study approach included an analysis of the four stations to learn more about their programmes and digital existence. Comparative approaches have become popular in media studies as they allow for the exploration of the same issue in various mediums. This is important to evaluate the scope and relevance of various phenomena (Esser & Hanitzsch, 2012), thus offering valuable insights into the studied phenomena. Radio content analysis focused on analysing the stations' content, posts images, and videos to understand more about their programming and levels of user participation. This involved listening to some of their live streams and broadcasts, which are readily accessible on their websites and social media platforms. By listening to the programmes, we learned more about the language used and the level of audience targeted in various shows. Our analysis focused on the stations spanning the last decade. We did not use interviews in the research as our focus was mainly on the stations' digital presence, programming and interaction with audiences.

## Discussion

The research demonstrated that the four stations under study are appropriating digital technologies in their day-to-day work. These platforms thus provide a platform for promoting participation of listeners interested in the stations who build communities of interest online around the shows/programmes. We noted that there is a convergence of radio, television and the internet, which enables multi-platform production of content which is distributed across networks. This convergence enables the stations to promote their reach to audiences dispersed across time and geography, evidenced through various communities on Facebook, Twitter, Instagram and TikTok and communities/listeners formed around specific shows and programmes. Mobile phones are also used in the process as listeners used them to participate in live programmes, through text messaging, comments, likes, shares and phone-ins and in consuming the programming of the various stations. The stations' existence on social media platforms such as Facebook, Twitter, Instagram and TikTok enables them to make stronger connections with their audiences. We have grouped our findings based on themes that cut across the four stations as we note more similarities in terms of adaptation and connection with audiences than differences.

### *A Strong Digital Presence and Creation of Online Communities*

All four stations have a strong digital presence on various platforms, including websites interconnected with social media platforms such as Facebook, Twitter, YouTube, TikTok and Instagram. Ukhozi FM has a strong presence on social media, with more than 700,000 followers on its Facebook page and more than seven million listeners. Facebook boasts of a combination of video, text and images, showing how the station has taken advantage of the platform's digital capabilities to interact with the target audience in various ways. Programming combines audio, video, image and text capabilities which audiences consume within their contexts. Similarly, 5FM has a strong social media presence, enabling it to connect with its youthful target audience or digital natives. 5FM's youth focus is made visible through the diverse youth images that dominate the station's website and social media platforms. It can be argued that these platforms have enabled the creation of communities that bring together users with similar interests bound by common concerns and also communities formed around specific social platforms. We argue that the stations create an overlap of local, national and transnational communities enabled by social media (Gunner et al., 2011).

In addition, Radio 947 has a significant digital footprint. The station's website presents all the critical information about contact details (postal code and address, email, telephone/mobile numbers for calls, SMS, fax and advertising), shows, podcasts, competitions, features and presenter descriptions. The website also has news articles with accompanying images and links to the station's social media accounts on Facebook, YouTube, Instagram and Facebook, as well as WhatsApp contacts. In addition, the website has an embedded "Listen to 947" play button that redirects to the live stream and two links to the station's podcasts on Apple Podcasts or Google Podcasts. The website also has links to the channel's mobile application, which is compatible with Android and Apple iOS. The station strongly emphasises convergence, as it offers text, images, video and audio all in one, like the other stations. The integration of digital tools on one platform thus shows how radio production in the present context has become interconnected or networked (Castells, 2010, 2013). Social networking sites have transformed in this context and become information networks in which the presenters, producers and audiences of the various stations produce and exchange information in real time (Castells, 2001).

Radio 947's YouTube channel started in 2009 and has close to 50,000 subscribers, with more than 11 million total channel views (June 2022 statistics). The channel brings the video aspect to radio broadcasting by providing snippets of in-studio broadcast experiences and special interviews with prominent people such as Dr John Kani, Noxolo Grootboom and Kelly Rowland. The Instagram account has 81,800 followers and close to 4000 posts (June 2022 statistics). These statistics indicate that there is relatively high following with audiences via this platform with snippets of the shows, posters of anticipated programmes, station event advertisements and competitions. The station's Facebook account has 306,920 followers and 276,291 likes and was started in 2009. It also provides snippets of shows and in-studio artist performances, news updates, competitions and a Top 40 countdown, and acts as an engagement tool for questions and comments during live broadcasts, which promotes interactivity between the station and the audiences and among the audiences themselves. Finally, the station's Twitter account has the most followers out of all the platforms at 422,900. The station joined Twitter in 2009 and has been used as a platform to advertise the station's content, questions, competitions and events. It has also been used for affiliated adverts from companies like Cell C, NutriDay and Spur. However, though there are many followers, there is very little audience feedback and engagement with actual tweets, thus showing that audience numbers do not equate to the extension of the public sphere.

The Kaya FM website also contains all the station's contact information (postal code and address, email, telephone/mobile numbers for calls and advertising) and details about shows and presenters. In addition, the website has links to embedded podcasts, competitions and news articles illustrated by images. The station's social media accounts on Facebook, YouTube, Instagram,Facebook and WhatsApp are also directly linked to the website. The live stream for the station is also linked through an embedded "listen live" button. The Kaya FM mobile application is linked to Google Play, Apple Store and App Gallery on the website's homepage.

When considering the digital presence of Kaya FM, we noted that the YouTube channel of the station, which was started in 2011, is also known as Kaya TV—this is indicative of the station's deliberate effort to create content that emphasises cross-platform engagement. With 176,000 subscribers, the station has various video snippets of in-studio broadcasts, external interviews and event coverage. Kaya FM has nearly 41 million total channel views, which points to the relevance of the content that it has shared for the past decade. Some of the videos shared emphasise the station's close relations with corporate advertisers such as CNA and KFC, as seen with the Back to School with Sandton City and the Kaya FM Family Fun Day with KFC. The connection with advertisers shows the channel's efforts to make money online. The channel also has aspects of political and current affairs discussions in isiZulu, which is the language of some of its listeners, on programmes such as Skhumba Talks, in which Skhumba Hlophe discusses Post Xenophobia and, in another episode, about lockdown effects on ordinary people. The channel also has a Kaya FM TV news coverage segment in which the station takes advantage of audio news-gathering techniques and video footage to communicate its news stories. A Kaya TV talk show called "The Red Couch with Mbali Dhlamini" also uses isiZulu to engage with some of the guests and thus shows that audiences and guest language preferences can shape how content is presented in line with the notion of the social constructivist approach on considering the context. The Kaya FM YouTube channel content shows that the station strategically uses this platform to provide its audiences with a radio and television experience. Thus, we note the extension of the digital public sphere and the creation of an online community around an issue of public interest.

The station also uses Facebook and Instagram to provide snippets of shows and in-studio artist performances and interviews, news updates, and competitions, and acts as an engagement tool for questions and comments during live broadcasts. The station's Facebook account has 423,284 followers and approximately 330,000 likes. The Instagram account has approximately 74,000 followers and 4357 posts, and there is relatively high interaction with audiences via this platform with short clips of live and previously broadcast shows, posters of anticipated shows, station event advertisements, and competitions. The station's Twitter account has the most followers out of all the platforms, that is, 714,400 followers. The station joined Twitter in 2009, using it as a platform to advertise the station's content, questions, competitions and events. It has also been used for affiliated adverts from companies like Mpumalanga Tourism, McCafe and Spar. Like 947, the Twitter account has many followers, but there is very little audience feedback and engagement with the actual posts. When looking at the lack of strong engagement from the audience side on the Twitter account, it may be argued that though audiences may have easy access to a strong network of communication, they can decide not to engage strongly, which points to agency in communicative processes. There is also the possibility that the Twitter posts have been communicated in a one-way format that does not encourage audience engagement and feedback, thus showing limitations of this new public space.

### *An Interplay of Culture and Community Online*

Broadcasting in IsiZulu, Ukhozi FM and, in some cases, Kaya FM signify an interplay of culture and community. Gunner (2017) argues that UKhozi FM forges intimacy and community with its audiences, allowing urban and rural listeners to navigate post-apartheid Zulu identity. Using a vernacular language allows various language slants, such as slang, to be incorporated and connect to "EKasi"—township home/community life. We also note a combination of English and IsiZulu on the broadcasts to cater for the diversity of culture of the audiences of the station. The promotion and dissemination of Zulu culture through Ukhozi FM is an important role played by the station, like the roles played by indigenous language media within the African context (Salawu, 2013). Through the streaming of isiZulu programmes such as #KubambaEzingelayo or

#NciphaBugebengu, as seen on its social media platforms, the station is disseminating Zulu culture to local and transnational audiences, thus reinforcing the role of the internet and digital platforms in the information dissemination and connection of geographically dispersed audiences.

### *Co-Creation of Programmes with Users*

Radio content analysis of the Ukhozi, 5FM, 947 and Kaya FM stations shows that they engage in co-creation/co-production with audiences. This is evident when the stations encourage the audiences to submit their music requests or voice notes about topical issues through WhatsApp or any other social media to be broadcast. This, we argue, not only allows for the development of local voices but also enables the stations to be contemporary public spheres and their audiences to engage in the co-creation of content they value. This is supported by Tettey et al. (2011, p. 24), who argues that the digital era is characterised by a "convergence of platforms that lends itself to multiple sources of information production and viewpoints, as well as the endless expanse of recipients".

### *A Convergence of Radio, TV and Social Media*

An analysis of Ukhozi, 5FM, 947 and Kaya FM shows a growing move towards greater convergence of radio, TV and social media in how the stations operate. It is important to note that all four stations currently offer five or more ways to access their content, that is, through the frequency finder, web browser, mobile radio and mobile apps (via Android OS or iOS) such as Facebook, Instagram, Twitter, TikTok and DSTV. The radio broadcasts demonstrate an overlap of the visual, audio and text. While listening to the radio, a listener also gets a full view of what is happening within the studio, thus engaging in television viewing simultaneously. The use of various digital platforms, which are often internet-enabled, presents increased communicative potential, which "is cheap, easily accessible, offers infinite channels, overcomes barriers of time and space and is not source restrictive" (Davis, 2010).

### *Radio as a Public Sphere*

Radio is often associated with the ability to form close relationships with audiences as content is produced to integrate easily into audiences' daily lives unobtrusively. As noted by Bosch (2022, n.p.):

> Despite the continued popularity of vernacular radio, English-language talk radio stations and shows still attract African language speakers who frequently phone in and participate. This could be linked to the dominance of English-language media in South Africa and the fact that English media spaces are also often dominant. In other words, despite the range of vernacular options, English stations are perceived as being sites of the public sphere and attract debate and conversation between a diverse range of South Africans. Regardless of language, talk radio shows are booming with vibrant conversations, highlighting the important role of radio as a space to bring together geographically diverse South Africans to debate matters of social and political importance.

The above quote demonstrates the important role of talk radio and how digital platforms facilitate debate and conversations, thus constituting an online public sphere. However, while the chapter has shown how the four stations have adapted in the digital era, we note that the digital divide curtails full participation online as it excludes those without access. Also, the lack of participation and minimal feedback on some of the posts on the social media platforms shows that social media is not wholly participative, and this could be a result of several social, political and economic factors. Not everyone has the capacity to participate on the online platforms. As Papacharissi (2002) argues, the creation of a digital public space does not mean a new public sphere. The digital divide still has a significant impact on how information is accessed and shared (Mutsvairo & Ragnedda, 2019). Digital activity remains limited to those with smartphones and internet to access these online platforms. It is also limited to those who are literate in the English language, which is the language that dominates websites and social media platforms.

## Conclusion

This chapter has argued that the digital era provides radio broadcasting organisations with agency and enables them to take advantage of the capabilities offered by new technologies to strengthen their offerings and build stronger relationships with audiences. We note how the creation of a strong presence online has enabled all the stations to be interconnected with local and global audiences who form communities offline and online. This is because of radio and internet's ability to transcend borders. The social constructivism theory has been useful in helping us analyse how the four radio stations have appropriated technological tools within the South African context. This study can be expanded to include the opinions and experiences of individuals who use these digital platforms, such as producers, presenters and audiences.

## References


Bosch, T. E. (2006). Radio as an Instrument of Protest: The History of Bush Radio. *Journal of radio studies, 13*(2), 249–265.

Bosch, T. (2010). Digital Journalism and Online Public Spheres in South Africa. *Communication, 36*(2), 265–275. https://doi.org/10.1080/02500167.2010.485374

Bosch, T. (2014). Social Media and Community Radio Journalism in South Africa. *Digital Journalism, 2*(1), 29–43. https://doi.org/10.1080/21670811.2013.850199

Bosch, T. (2022). Radio is Thriving in South Africa: 80% are Tuning In. https://theconversation.com/radio-is-thriving-in-south-africa-80-are-tuning-in-176846

Briefly. (2021). List of the Top 20 South African Radio Stations and Their Frequencies. https://briefly.co.za/107472-list-top-20-south-african-radio-stations-frequencies.html

Broadcast Research Council of South Africa (BRC). (2021). BRC Radio Audience Measurement Listenership Report Apr 21–March 22.

Castells, M. (2001). *Challenges of Globalisation: South African Debates with Manuel Castells* (Vol. 4). Pearson South Africa.

Castells, M. (2010). *The Rise of the Network Society* (2nd ed.). Blackwell Publishing.

Castells, M. (2013). *Communication Power*. Oxford University Press (UK).

Chiumbu, S. H., & Ligaga, D. (2013). 'Communities of Strangerhoods?': Internet, Mobile Phones and the Changing Nature of Radio Cultures in South Africa. *Telematics and Informatics, 30*(3), 242–251.

Coyne, R. (2010). *The Tuning of Place: Sociable Spaces and Pervasive Digital Media*. MIT Press.

Davis, A. (2010). Media and Politics. In J. P. Curran (Ed.), *Mass Media and Society* (5th ed.). Bloomsbury USA.

Esser, F., & Hanitzsch, T. (1969 [2012]). *Handbook of Comparative Communication Research*. Routledge, London; New York.

Evans, M. (2022). Radio has a Rich History as a Weapon of the Liberation Struggle in Southern Africa. https://theconversation.com/radio-has-a-rich-history-as-a-weapon-of-the-liberation-struggle-in-southern-africa-177041
Fombad, M. C., & Jiyane, G. V. (2019). The role of community radios in information dissemination to rural women in South Africa. *Journal of Librarianship and Information Science, 51*(1), 47–58.
García-Avilés, J. A. (2021). Journalism as Usual? Managing Disruption in Virtual Newsrooms during the COVID-19 Crisis. *Digital Journalism, 9*(9), 1239–1260. https://doi.org/10.1080/21670811.2021.1942112
Gunner, L. (2000). Wrestling with the Present, Beckoning to the Past: Contemporary Zulu Radio Drama. *Journal of Southern African Studies, 26*(2), 223–237.
Gunner, L. (2017). The Man from Where? Ukhozi FM and New Identities on Radio in South Africa. *Journal of African Media Studies, 9*(1), 61–75.
Gunner, L., Ligaga, D., & Moyo, D. (2011, 2012, 2010). *Radio in Africa: Publics, Cultures, Communities.* Wits University Press.
How, T. T. (2010). *Some Questions on Technological and Social Determinism.* Paper presented at the 19th Annual Conference of the Asian Media Information and Communication Centre, Singapore. http://www.spp.nus.edu.sg
Mabweazara, H. M. (2015). African Journalism in the 'Digital Era': Charting a Research Agenda. *African Journalism Studies, 36*(1), 11–17.
Matrix in South Africa During the National Arts Festival. https://doi.org/10.1177/1354856508097020
Matsilele, T., Makwambeni, B., Nkoala, S., & Bulani, G. J. (2023). Youth Audiences and Social Media Integration in Community Radio Stations in South Africa: A Case Study of Zibonele FM and Bush Radio. In S. Tsarwe & S. Chiumbu (Eds.), *Converged Radio Youth and Urbanity in Africa. Emerging Trends and Perspectives* (pp. 123–153). Springer International Publishing.
Mhlambi, T. N. (2008). The Early Years of Black Radio Broadcasting in South Africa: A Critical Reflection on the Making of Ukhozi FM. Unpublished master's thesis, University of Cape Town.
Mtimde, L. (2003). Regulatory Options to Guarantee More Democratic Media and Information Technologies. *Paper delivered at OURMedia III–Strengthening Citizens' Communication at Barranquilla*. Colombia.
Mutsvairo, B. (2019). A New Dawn for the "Developing" World? Probing Data Journalism in Non-Western Societies. In B. Mutsvairo, S. Bebawi, & E. Borges Rey (Eds.), *Data Journalism in the Global South.* Palgrave Macmillan.
Mutsvairo, B., & Ragnedda, M. (2019). *Mapping Digital Divide in Africa: A Mediated Analysis.* Amsterdam University Press. https://doi.org/10.2307/j.ctvh4zj72
Ndlovu, M., & Sibanda, M. N. (2021). Digital Technologies and the Changing Journalism Cultures in Zimbabwe: Examining the Lived Experiences of Journalists Covering the COVID-19 Pandemic. *Digital Journalism.* https://doi.org/10.1080/21670811.2021.1976065

Ndzinisa, N. T., Lunga, C. M., & Ndlovu, M. (2021). News in the Digital Age: A Case Study of CITE as a Digital Public Sphere in Zimbabwe. *African Journalism Studies, 42*(4), 46–64. https://doi.org/10.1080/23743670.2022.2028647

Nkoala, S. (2023). Radio in South Africa Turns 100—And Collides with Podcasting and Streaming. https://theconversation.com/radio-in-south-africa-turns-100-and-collides-with-podcasting-and-streaming-198021

Nkoala, S., & Matsilele, T. (2021). In Whose Interest is it Anyway?: An Analysis of the Content, Discourse, and Programming Strategies of South African Business Radio Shows. In *Radio, Public Life and Citizen Deliberation in South Africa*. Routledge.

Nkosi, L. (2014). Black Women in Post-apartheid South Africa, Nation-Building and Radio: The Case of Ukhozi FM. Unpublished master's thesis, University of Cape Town.

O'Gorman, N. (2018). Media Technologies in Communication and Critical Cultural Studies. https://doi.org/10.1093/acrefore/9780190228613.013.548

Papacharissi, Z. (2002). The Virtual Sphere: The Internet as a Public Sphere. *New Media & Society, 4*(1), 9–27.

Salawu, A. (2013). Access, local language and public sphere: Phone-in R/TV programme as a platform for citizen journalism in a Nigerian election. *Telematics and Informatics, 30*(3), 267–274.

Tettey, W. J., Mano, W., Bosch, T., Davis, S. R., Wolters, S., Lekgoathi, S. P., Straus, S., Chibita, M. B., Moorman, M. J., Schulz, D. E., & Coplan, D. (2011). *Radio in Africa: Publics, Cultures, Communities*. Wits University Press.

Tyali, S. M. (2020). Re-reading the Propaganda and Counter-Propaganda History of South Africa: On the African National Congress' (ANC) Anti-Apartheid Radio Freedom. *Critical Arts, 34*(4), 61–75.

Verdone, R., & Zanella, A. (Eds.). (2012). *Pervasive Mobile and Ambient Wireless Communications: COST Action 2100*. Springer Science & Business Media.

Verweij, P. (2009). Making Convergence Work in the Newsroom: A Case Study of Convergence of Print, Radio, Television, and Online Newsrooms at the African Media. *Convergence, 15*(1), 75–87.

Wake, A., & Bahfen, N. (2016). Redefining Radio: Implications for Journalism Education in an Era of Digital Audio Storytelling. *Radio Journal, 14*(2), 231–242.

## Websites

5FM. http://www.5fm.co.za/sabc/home/5fm/

Kaya FM. https://www.kaya959.co.za/

Radio 947. http://www.947.co.za/

Thebe Investment Corporation. http://www.thebe.co.za

Ukhozi FM. http://www.ukhozifm.co.za/sabc/home/ukhozifm

CHAPTER 9

# Emergence of Indigenous Language Podcasts in South Africa: Amplifying the Agency of Indigenous Language Speakers in the Digital Public Sphere

*Sisanda Nkoala*

## Introduction and Literature Review

Podcasting has experienced remarkable growth in South Africa in recent years, reflecting an increasing interest and demand for digital audio content (van der Merwe, 2021). While specific listenership figures are challenging to obtain, the surge in advertising spend by companies demonstrates the growing appeal and potential of podcasts in the country (PwC, 2022, p. 12). The Africa Entertainment and Media Outlook 2022 report published by PwC in October 2022 reveals that podcast advertising revenue witnessed a significant growth rate of 30.4% in 2020 (PwC, 2022, p. 12). In terms of platforms, Spotify emerges as the most popular platform

S. Nkoala (✉)
Cape Peninsula University of Technology, Cape Town, South Africa



S. Nkoala, G. Motsaathebe (eds.), *100 Years of Radio in South Africa, Volume 2*, https://doi.org/10.1007/978-3-031-40706-2_9

among South African podcast consumers, followed by Apple Podcasts (PwC, 2022, p. 12). Compared to radio, a feature that enhances the parasocial intimacy in podcasts is the fundamental opt-in orientation of these digital texts (Bottomley 2015). Unlike terrestrial radio, where one is exposed to whatever content happens to be on-air at the time of listening, when it comes to podcasts, audiences need to not only find the content, they then need to subscribe. They then choose when they listen, unlike radio consumption, which is directed by a set schedule. While this flexibility of consumption is a crucial feature, there are generally accepted conventions regarding productions, which include the expectation that a particular podcast will publish an episode at regular intervals and that the format of the programme will have some features that characterise a particular series.

This makes their uptake in the African context noteworthy because although issues of technological access and affordability still limit the creators and consumers of these digital texts, their "ability to engage communities is enhanced by some of the defining characteristics of podcast aesthetics, namely their parasocial intimacy—that is, the tendency for listeners to think of their favourite podcast hosts as 'friends in their ears'" (McGregor, 2022).

However, when examining the top podcast charts in South Africa, it becomes evident that the most widely listened to podcasts are predominantly in English and often produced by non-South African entities. Figure 9.1, sourced from Chartable on 7 July 2023, provides a snapshot of the top podcasts in South Africa on the popular streaming platform Spotify. The chart showcases the disparity between the representation of local content and the dominance of English language and non-South African podcasts.

The chart reveals that out of the top 10 podcasts, only three are produced by South Africans: *Podcast and Chill with MacG*, *VT Podcast "Ideas That Matter"*, and *Wisdom and Wellness with Mpoomy Ledwaba*. All three are in English. This contrasts with the radio landscape, where indigenous language radio stations attract a more extensive listenership due to their ability to cater to the linguistic preferences of indigenous language speakers, as shown in Fig. 9.2. This disparity raises questions about the visibility and recognition of local voices within the podcasting landscape. Despite the linguistic diversity and rich cultural heritage in South Africa, indigenous language podcasts struggle to break into the mainstream and achieve

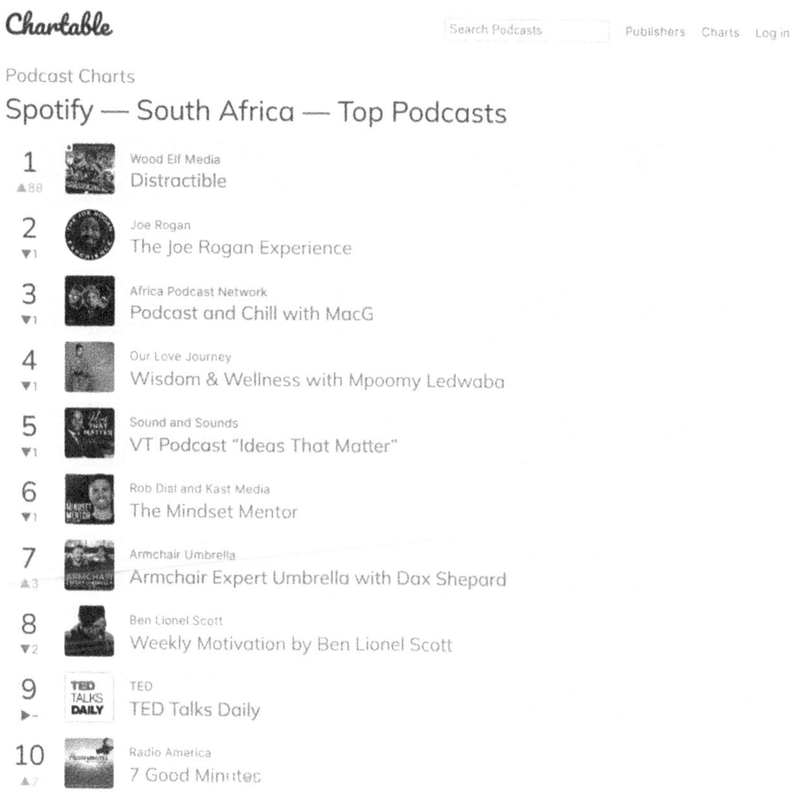


**Fig. 9.1** https://chartable.com/charts/spotify/south-africa-top-podcasts (12 July 2023)

similar levels of popularity (Africa Podcast 2021). The dominance of English language and non-South African podcasts in the chart highlights the need for increased representation of local content and diverse voices. Indigenous language podcasts have the potential to capture the narratives, experiences, and cultural perspectives that are often overlooked or underrepresented in mainstream media. The podcasting landscape can foster a more inclusive and representative environment by providing a platform for local content creators and amplifying indigenous languages (Florini 2020).

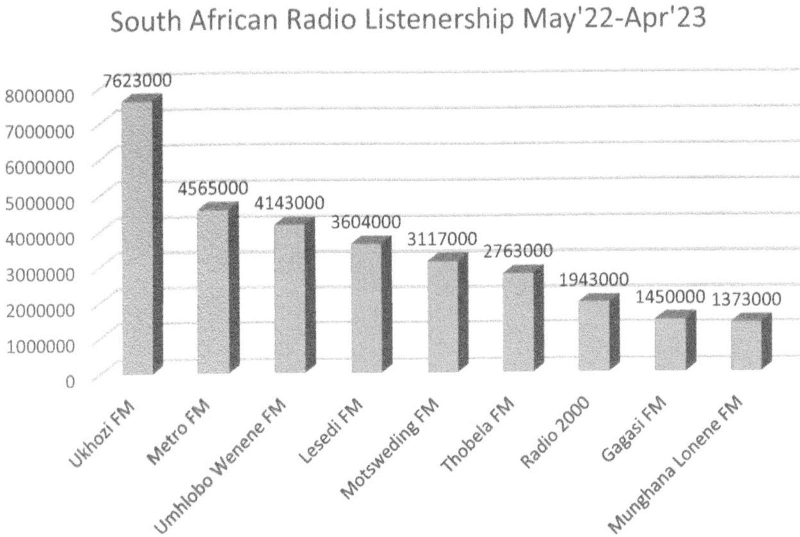


**Fig. 9.2** South African radio listener figures. (Source: https://brcsa.org.za/rams-amplify-radio-listenership-report-may22-apr23/)

The place of podcasts in South Africa must be understood within the context of the rich 100 years of radio in the country, where oral media has been a central platform for developing linguistic and cultural identities. When radio first arrived in South Africa, it was not meant to include broadcasts to indigenous language speakers. It was only as part of efforts to gain the support of the majority black population during the Second World War that black people were even thought of as a possible audience (Mhlambi, 2019, p. 213). These broadcasts, initially brief, gradually increased in duration. It became evident that radio could be harnessed to shape black South Africans' perspectives based on the white authorities' interests (Mhlambi, 2019, p. 213). Once radio was opened to indigenous language speakers, both as an audience and content creators, it became a powerful tool for expressing their voices and challenging societal silencing (Gunner et al., 2011). Despite its initial design as a means of colonial propaganda, radio demonstrated a multifaceted nature and the ability to operate within the realm of cultural knowledge, giving rise to new and often unruly publics (Gunner et al., 2011, p. 6). Language played a pivotal role in the effectiveness of radio, enabling it to become a platform for cultural expression and the development of indigenous language communities.

The literature on radio and indigenous language-speaking audiences in South Africa further supports the notion of radio as a vital medium for engaging with and serving the needs of indigenous language speakers (Onyenankeya & Salawu, 2019; Gunner et al., 2011; Lekgoathi, 2009). The scholarship shows that Indigenous language radio stations have consistently attracted more listeners by catering to the linguistic preferences of their respective communities. These stations provide a space for indigenous language speakers to hear content in their native tongues, fostering a sense of cultural belonging and identity (Onyenankeya & Salawu, 2019). Moreover, indigenous language radio has played a significant role in promoting local music, arts, and cultural expression, giving a voice to artists and communities that have been marginalised within the broader media landscape (Odine, 2013). In recent years, the rise of indigenous language podcasts has marked a significant development within the South African media landscape (Alegi, 2012). These podcasts provide a platform for indigenous language speakers to reclaim their voices in a digital media landscape dominated by English, share their stories, and address topics relevant to their communities. By leveraging the affordances of the digital public sphere, indigenous language podcasters can reach broader audiences and challenge the dominance of English language content in the podcasting sphere (Florini, 2020; Nagy & Neff, 2015).

These podcasts serve as platforms for storytelling, cultural expression, and the dissemination of knowledge within indigenous language communities (Abel, 2015; McHugh, 2016).

By analysing *Epokothweni*'s production practices, including its use of social media platforms and digital distribution channels like Spotify, Apple Podcasts, and Google Podcasts, the chapter builds on the scholarship on audio media in this context and examines how the podcast leverages the digital public sphere to engage with its audience and further provide socioculturally relevant content (Nonkenge, 2021, 2022a). The digital public sphere encompasses online spaces where individuals engage in public discourse, providing opportunities for marginalised voices to be heard and meaningful conversations to occur. *Epokothweni*'s success in reaching black South African audiences underscores the need for platforms prioritising indigenous languages and challenges the dominance of English language content in the podcasting landscape.

## Theoretical Framework

The theoretical framework for this study is the digital public sphere, which conceptualises online spaces as platforms for public discourse and civic engagement. Habermas defined the public sphere as a space where private individuals gather to discuss matters of general interest, with guaranteed access and freedom of expression (Habermas, 1989). He described it as "A realm of our social life in which something approaching public opinion can be formed. Access is guaranteed to all citizens. A portion of the public sphere comes into being in every conversation in which private individuals assemble to form a public body… Citizens behave as a public body when they confer in an unrestricted fashion—that is with the guarantee of freedom of assembly and association and the freedom to express and publish their opinions—about matters of general interest" (Habermas, 1989, pp. 73–74). The digital public sphere theory builds upon this conception by acknowledging the transformation of the public sphere in the digital age. It recognises that the virtual sociality of platforms like the internet refracts the public sphere, introducing elements such as on-demand listening, cost-effectiveness, and user-generated content (Jenkins & Myers, 2022; Swiatek, 2018).

The digital public sphere differs from Habermas's original notion by creating alternative spaces for subaltern counter-publics. These counter-publics utilise digital tools to question the established consensus in mainstream discourse (Castells, 2008). Subordinated groups, such as indigenous language speakers, can utilise the affordances provided by the digital public sphere to become public opinion leaders. The digital public sphere transcends traditional media boundaries, allowing diverse voices to participate in conversations that shape public opinion and challenge existing power structures. Within the digital public sphere, individuals and communities can express themselves, share experiences, and advocate for sociocultural change (Florini, 2020; Nagy & Neff, 2015).

The digital public sphere allows marginalised voices to be heard, fosters dialogue, and promotes sociocultural relevance, empowering indigenous language speakers to articulate their experiences and shape public discourse (Giroux, 2011; Fox et al., 2020). By adopting the framework of the digital public sphere framework, this study examines how indigenous language podcasts, such as *Epokothweni*, contribute to forming a more inclusive and participatory media environment in South Africa.

The argument advanced in this chapter is that the digital nature of these podcasts allows the producers of these texts leeway to create indigenous language communities online and on-air by drawing on discursive practices such as storytelling and conversations within the digital public sphere. Fox et al. (2020, p. 7) are of the view that podcasts produced for black audiences are "particularly representative of online media's capacity to usher new voices into the conversation, forging alliances across national and ethnic boundaries, and protecting those public spheres where the formative culture necessary for creating educated and informed citizens can develop".

## Methodology

This qualitative study examines the phenomenon of indigenous language podcasts in South Africa, using *Epokothweni* as a case study. *Epokothweni*, hosted by Babalwa Nonkenge, focuses on personal financial stewardship in isiXhosa. The podcast has gained recognition in the South African media landscape and is on various platforms, including social media, YouTube, Facebook, WhatsApp, Spotify, Apple Podcasts, and Google Podcasts.

The study uses a case study approach to provide in-depth and detailed insights into a specific phenomenon or context of indigenous language podcasting in South Africa by considering *Epokothweni* (Crilly, 2019). This study's data collection and analysis approach was audio content analysis. This methodology systematically examines and interprets audio data to extract meaningful insights (Lindgren et al., 2020). In studying podcasts, audio content analysis involves listening to podcast episodes, transcribing spoken words, identifying key themes or topics, analysing tone and emotions conveyed through voice, examining nonverbal cues, and exploring the overall structure of the audio content. This study focused on discourse and language to understand the linguistic, cultural, and discursive dimensions present in the *Epokothweni* podcast. The researcher listened to a series of episodes, purposefully sampling portions that demonstrate the use of radio storytelling practices and their socio-cultural relevance. These sampled portions were transcribed and analysed to consider the audio recordings' discourse, language, multimodal features, storytelling techniques, cultural expressions, and contextual nuances.

*Epokothweni* is hosted and produced by financial expert Babalwa Nonkenge. The podcast was started in 2021 as a platform "to talk all things personal financial stewardship in [her] mother tongue isiXhosa" (Nonkenge, 2021). The Pan South African Language Board awarded it the 2022 technology and business award. The podcast has a presence on social media using the handle @epokothweni on Twitter and Instagram. It is also available on YouTube, Facebook, and WhatsApp. The podcast is published on various platforms, including Spotify, Apple, and Google (Nonkenge, 2022a).

## Discussion

As a medium that is often lauded for its ability to reach millions while maintaining a sense of intimacy through a sense of personal presence, one of radio's enduring legacies in South Africa has been in how, despite being a mass medium, it can transfix the audience through references to the individual listener. How South Africans consume radio as a type of background companion in their everyday activities has shaped production practices such as scheduling of content to align with what the producers of this content imagine the public to be doing at a particular time and choice in presenters based on assumptions of who the imagined audience is at any given time. One would assume that such production practices would not translate into podcasting since this content is consumed using an on-demand approach that is not informed by audiences tuning in at a specific time or place. However, with *Epokothweni*, we hear traces of that in how, in her opening, the host encourages the listener to grab a beverage and a notebook and something to write with as the podcast gets started. An example of this is in episode 30, where she says:

> *Ndiyathembe uzakuthi zava. Thatha into yakho yokuphunga nje ngesiqhelo, uthathe nesibhilivana sakho apho apho ubhala khona ukwazi ukwenza ii-notes zakho* [I trust you will relax. Grab a hot beverage as usual, and grab your notebook that you write in so that you can take your notes].

Here she refers to one of the enduring ways in which radio has been consumed in the South African context, namely a domestic context where one can grab a beverage, relax, and take notes. This imagined context on the part of the producer of *Epokothweni* informs the rest of the approach taken in the podcast, which has a very relaxed conversational tone. If one considers the programming clock structure that has characterized radio in South Africa for decades, where we have a morning breakfast show that typically airs from 06:00 to 09:00, a mid-morning show from 09:00 to 12:00, a lunchtime show from 12:00 to 15:00, and an afternoon drive time show from 15:00 to 18:00, one could describe the content of this podcast as falling into the mid-morning or lunchtime programming slots, or perhaps late from 20:00 onwards (Setiowati, 2018). These are the slots where people would most likely be able to take notes while drinking a beverage and listening to a podcast. Also, the suggestion for the listeners to take a notebook and make notes implies that the people listening are doing so in a context where they can give the content their undivided attention, so it is not someone listening while cooking or driving, for example. This is relevant in South Africa because the mid-morning, lunchtime, and late evening slots are typically associated with this kind of content.

Indigenous language podcasts are significant in the digital public sphere, where imagined audiences play a crucial role. The concept of imagined audiences, as used in this chapter, pertains to content producers' understanding and assumptions about their target listeners and the specific contexts in which they imagine them to interact with the content (Nelson, 2021). In the context of indigenous language podcasts and South African radio, this concept gains further significance as it delves into how content producers adapt their production practices to meet the needs and preferences of their envisioned audience within the digital public sphere.

In the case of *Epokothweni*, the host establishes an imagined context by encouraging listeners to grab a beverage and a notebook, reminiscent of the traditional radio experience in South Africa. This practice aligns with how radio has been consumed as a companion in everyday activities, creating a sense of intimacy between the broadcaster and the individual listener. Despite the on-demand nature of podcasting, *Epokothweni* maintains this sense of personal presence by referencing familiar contexts and activities, such as relaxing with a hot beverage and taking notes. The content on *Epokothweni* positions the podcast as engaging directly with listeners as friends and regularly addresses them as *izihlobo zasepokothweni* (friends of *Epokothweni*), as well as encouraging listeners to make contact with the host through the platform's other social media channels. Even when the host has a guest or guests she is interviewing, the references to the "friends of *Epokothweni*" position the engagement as a three-way conversation.

By examining the imagined audience of *Epokothweni*, we observe the production practices that align with specific time slots traditionally associated with radio programming in South Africa. The mid-morning, lunchtime, and late evening slots are typically when individuals can engage with content more attentively, making them suitable for content that encourages note-taking and active listening. By aligning its production practices with specific time slots traditionally associated with radio programming, *Epokothweni* demonstrates a conscious adaptation of established storytelling platforms to the digital realm. This adaptation highlights the podcast's recognition of the audience's preferences and habits and their availability to engage with content more attentively during specific periods. Moreover, this underscores the significance of understanding the historical context of radio in South Africa and its influence on the production strategies employed by indigenous language podcasts. By leveraging the materiality and functionality of radio, *Epokothweni* harnesses the familiarity and trust associated with this established medium, thus enhancing its connection with the imagined audience within the digital public sphere.

Furthermore, the tone of the conversation in *Epokothweni* reflects an educative approach, with the host leveraging her background as a commerce professional to impart financial knowledge to the audience. This positioning as an educator and using metaphors and storytelling techniques create a sense of trust and expertise, aligning with the concept of counter-elites within the digital public sphere. Counter-elites refer to individuals or groups who challenge the dominant power structures and discourses (Badr, 2021). In the context of *Epokothweni*, the host assumes the role of a counter-elite by offering financial insights and advice in indigenous languages, which have been historically marginalised in mainstream financial discourse. This positions *Epokothweni* as a platform that empowers marginalized communities, such as indigenous language speakers, by providing relevant content in their language.

When the presenter speaks to the audience, she addresses them as *izihlobo zasepokothweni*, fostering a sense of community and participation. *Epokothweni*'s emphasis on community-building within the digital public sphere aligns with the concept of participatory culture. Participatory culture, as defined by Jenkins, refers to a culture in which individuals actively create, circulate, and interpret media content

(Jenkins, 2019). Through its educative approach and the inclusion of the audience as active participants, *Epokothweni* cultivates a participatory culture among its listeners. By addressing the audience as friends and encouraging them to interact through social media channels, *Epokothweni* creates a space for dialogue and knowledge-sharing, allowing the audience to contribute to the ongoing conversation and shape the direction of future episodes. This participatory approach within the digital public sphere expands the possibilities for indigenous language speakers to actively participate in public discourse and challenge the dominant narratives that have excluded them in the past.

*Epokothweni*, as a podcast primarily sharing content in Xhosa, employs multilingual communication strategies that reflect the everyday manner in which South Africans communicate. These strategies include code-switching, translanguaging, and translation, where different registers of Xhosa and English are used to cater to a diverse audience. In episode 46 of *Epokothweni*, the host talks to a guest about different financial products, such as unit trusts.

> *Namhlanje sizakuncokola ngalendawo ibonakalayo. Siye sayifanisa ne-container okanye isingxobo apho yenzeka khona yonke le-action* [Today, we are going to talk about the aspect that is visible. We likened it to a container, or a container, where all of this action happens].

The host does not mention that she is sharing a translation of the word container when she says *okanye isinxgobo.* She just says the two words. Xhosa speakers will know from listening to the context that she is engaged in translating the English term into Xhosa and come to know of a term they might not have previously known. This approach acknowledges that many South Africans are multilingual, speaking an indigenous language as their mother tongue and English as the language of business and scientific engagement. Using translation techniques, *Epokothweni* bridges the gap between languages and gives listeners a deeper understanding of complex concepts. These translation practices demonstrate the podcast's commitment to accessibility and inclusivity, moving away from the purist and separatist approach that characterized indigenous radio content during apartheid.

The use of these translation techniques in this multilingual context is part of the everyday manner in which South Africans communicate, whereby many will speak an indigenous language as their mother tongue but also know how to speak English as the language of business and scientific engagement in this country, and is another example of how the podcast fosters a participatory culture in the digital public sphere. *Epokothweni* encourages listeners to actively engage with the podcast by utilising translation techniques and incorporating different languages, expanding their vocabulary and deepening their understanding of complex topics. The availability of social media platforms further facilitates real-time conversations between the audience and the podcast's creators, creating a multimedia ecosystem where engagement extends beyond the podcast episodes themselves. This participatory approach strengthens the sense of community, promotes knowledge exchange, and empowers listeners to actively shape the content and discussions within the digital public sphere.

*Epokothweni*'s multilingual production practices align with the concept of translanguaging, which emphasises the fluid and dynamic nature of language use. Translanguaging recognises that language boundaries are not fixed and encourages the blending and negotiation of languages for effective communication (Makalela, 2022). By actively incorporating multiple languages into the podcast, *Epokothweni* embraces the linguistic diversity of its audience and ensures that individuals from different language backgrounds can engage with the content. This approach challenges the monolingual assumptions often imposed by content producers and recognises the multilingualism of black South Africans. When audio broadcasts to black South Africans were initially introduced in the 1940s, the inclination was to stratify the audience according to ethnic and linguistic identity by way of stations that were intended to broadcast exclusively in specific languages, and the apartheid government had planned to use radio as an instrument to entrench further the ideology espoused by the system of separateness (Gunner et al., 2011). In the context of 100 years of radio in South Africa, *Epokothweni*'s departure from this purist and separatist approach is significant as it challenges the historical legacy of language-based separation and exclusion. By employing multilingualism and embracing translanguaging practices, *Epokothweni* positions itself as a platform that celebrates the linguistic diversity of South Africa's indigenous language speakers, fostering inclusivity and actively promoting engagement within the digital public sphere.

The representation of indigenous languages in podcasts like *Epokothweni* is significant because it addresses the historical marginalisation and underrepresentation of these languages in mainstream media.

By providing content in indigenous languages, *Epokothweni* empowers indigenous language speakers by offering them access to information and knowledge in their mother tongue. This matters because language is closely tied to cultural identity, and promoting indigenous languages fosters a sense of pride, cultural preservation, and self-expression. It challenges the dominant narratives perpetuating monolingualism and exclusion in public discourse, promoting linguistic diversity as a strength rather than a barrier. Secondly, adapting radio production practices within the podcasting context demonstrates indigenous language media's resilience and relevance in the digital age. Despite the on-demand nature of podcasts, *Epokothweni* maintains a personal and intimate connection with its audience by referencing familiar contexts and activities associated with radio consumption. This matters because it signifies the continuity of indigenous language media as a powerful means of communication and cultural preservation. It also highlights the ability of indigenous language podcasts to leverage the strengths of traditional media platforms while embracing the opportunities offered by digital technologies, ensuring the continued reach and impact of their content.

## Conclusion

The dominance of English language and non-South African podcasts in the top charts of the most popular podcasts in South Africa underscores the need for increased representation of local content and diverse voices within the podcasting landscape in South Africa. Despite the country's linguistic diversity and rich cultural heritage, indigenous language podcasts struggle to break into the mainstream and achieve comparable levels of popularity. This disparity highlights the importance of providing platforms prioritising indigenous languages and challenging the dominance of English language content. Indigenous language podcasts have the potential to capture narratives, experiences, and cultural perspectives that are often overlooked or underrepresented in mainstream media, thus fostering a more inclusive and representative environment.

Indigenous language podcasts like *Epokothweni* have the potential to foster linguistic diversity, challenge dominant narratives, and create a more inclusive and participatory media environment in South Africa. This chapter has explored the emergence of indigenous language podcasts in South

Africa and their role in amplifying the agency of indigenous language speakers within the media landscape. As a case study, it focused on *Epokothweni*, an award-winning personal finance podcast in isiXhosa. The study utilised the concept of the digital public sphere to examine how indigenous language podcasters use online spaces to address sociocultural issues and foster meaningful conversations within their communities.

The study argues that adapting radio production practices within the context of podcasting, as demonstrated by *Epokothweni*, showcases the resilience and relevance of indigenous language media in the digital age. Despite the on-demand nature of podcasts, *Epokothweni* maintains a personal and intimate connection with its audience by referencing familiar contexts and activities associated with radio consumption. This continuity of indigenous language media as a powerful means of communication and cultural preservation is significant.

It also shows how *Epokothweni* fosters participatory culture within the digital public sphere by creating a space for dialogue, knowledge-sharing, and community-building. Addressing its audience as "friends" and encouraging active participation through social media platforms, the podcast empowers marginalised communities to voice their perspectives, challenge dominant narratives, and contribute to public discourse. This inclusive approach challenges the historical legacy of exclusion and allows indigenous language speakers to shape media narratives and promote their cultural values actively. The multilingual production practices of *Epokothweni* further enhance inclusivity and accessibility, celebrating the linguistic diversity of South Africa's indigenous language speakers.

The scholarship on podcasts in South Africa has many areas of possible growth, including further research exploring the impact of indigenous language podcasts on audience engagement and empowerment, examining the sustainability and scalability of indigenous language podcasting initiatives, and investigating the role of digital platforms in facilitating the growth and reach of indigenous language podcasts. Future research could expand the investigation to include a broader range of indigenous language podcasts in South Africa and examine the dynamics of podcast consumption within diverse communities.

## References

Abel, J. (2015). *Out on the Wire: The Storytelling Secrets of the New Masters of Radio*. Crown.

Africa Podcast. (2021). 'Is this Mic On?': Exploring How Podcasting is Taking Root across Africa. *Baraza Media Lab*. Retrieved November 4, 2022, from https://barazalab.com/wp-content/uploads/2021/08/AfricaPodfest Report.pdf

Alegi, P. (2012). Podcasting the Past: Africa Past and Present and (South) African History in the Digital Age. *South African Historical Journal, 64*(2), 206–220.

Badr, H. (2021). Before the 'Arab Spring': How Challengers Pushed Counter-issues in Egypt's Hybrid Media System. *Media, War & Conflict, 14*(4), 522–541.

Bottomley, A. J. (2015). Podcasting: A Decade in the Life of a 'New' Audio Medium: Introduction. *Journal of Radio & Audio Media, 22*(2), 164–169. Retrieved October 18, 2022, from https://www.andrewjohnbottomley.com/s/Bottomley_JRAM-Introduction_2015.pdf

Castells, M. (2008). The New Public Sphere: Global Civil Society, Communication Networks, and Global Governance. *The Annals of the American Academy of Political and Social Science, 616*(1), 78–93.

Crilly, N. (2019). Creativity and Fixation in the Real World: A Literature Review of Case Study Research. *Design Studies, 64*, 154–168.

Florini, S. (2015). The Podcast 'Chitlin' Circuit': Black Podcasters, Alternative Media, and Audio Enclaves. *Journal of Radio & Audio Media, 22*(2), 209–219. Retrieved November 4, 2022, from https://doi.org/10.1080/19376529.2015.1083373

Florini, S. (2020). Podcasting Blackness. In L. K. Lopez (Ed.), *Race and Media: Critical Approaches*. New York University Press.

Fox, K., Dowling, D. O., & Miller, K. M. (2020). A Curriculum for Blackness: Podcasts as Discursive Cultural Guides, 2010–2020. *Journal of Radio & Audio Media, 27*(2), 298–318. Retrieved November 4, 2022, from https://doi.org/10.1080/19376529.2020.1801687

Giroux, H. A. (2011). The Crisis of Public Values in the Age of the New Media. *Critical Studies in Media Communication, 28*(1), 8–29. Retrieved December 5, 2022, from https://doi.org/10.1080/15295036.2011.544618

Gunner, L., Ligaga, D., & Moyo, D. M. (2011). Introduction: The Soundscapes of Radio in Africa. In L. Gunner, D. Ligaga, & D. M. Moyo (Eds.), *Radio in Africa: Publics, Cultures, Communities*. Wits University Press.

Habermas, J. (1989). *The public sphere: An inquiry into a category of Bourgeois society*. https://courses.ischool.berkeley.edu/i218/s10/JH-STPS.pdf.

Jenkins, B. M., & Myers, T. K. (2022). Digital Black Voices: Podcasting and the Black Public Sphere. *Digital Humanities Quarterly, 16*(3).

Jenkins, H. (2019). *Participatory Culture: Interviews.* John Wiley & Sons.

Lekgoathi, S. P. (2009). 'You are Listening to Radio Lebowa of the South African Broadcasting Corporation': Vernacular Radio, Bantustan Identity and Listenership, 1960–1994. *Journal of Southern African Studies, 35*(3), 575–594.

Lindgren, B. M., Lundman, B., & Graneheim, U. H. (2020). Abstraction and Interpretation During the Qualitative Content Analysis Process. *International Journal of Nursing Studies, 108*, 103632.

Makalela, L. (2022). *Not Eleven Languages: Translanguaging and South African Multilingualism in Concert* (Vol. 107). Walter de Gruyter GmbH & Co KG.

McGregor, H. (2022, June 20). Podcast Studies. *Oxford Research Encyclopedia of Literature.* Retrieved October 18, 2022, from https://oxfordre.com/literature/view/10.1093/acrefore/9780190201098.001.0001/acrefore-9780190201098-e-1338

McHugh, S. (2016). How Podcasting is Changing the Audio Storytelling Genre. *Radio Journal: International Studies in Broadcast & Audio Media, 14*(1), 65–82. Retrieved October 18, 2022, from https://ro.uow.edu.au/cgi/viewcontent.cgi?article=3366&context=lhapapers

Mhlambi, T. (2019). African Pioneer: KE Masinga and the Zulu 'Radio Voice' in the 1940s. *Journal of Radio & Audio Media, 26*(2), 210–230. Retrieved December 15, 2022, from https://doi.org/10.1080/19376529.2018.1468445

Nagy, P., & Neff, G. (2015). Imagined Affordance: Reconstructing a Keyword for Communication Theory. *Social Media + Society, 1*(2), 1–9. Retrieved October 18, 2022, from https://doi.org/10.1177/2056305115603385

Nelson, J. L. (2021). *Imagined Audiences: How Journalists Perceive and Pursue the Public.* Journalism and Pol Commun Unbo.

Nonkenge, B. (2021, June 23). *Epokothweni Instagram Account.* Retrieved October 17, 2022, from https://www.instagram.com/epokothweni/?hl=en

Nonkenge, B. (2022a). *Epokothweni.* [Online]. Retrieved October 17, 2022, from https://www.epokothweni.co.za/

Nonkenge, B. (2022b). *Epokothweni with Babalwa Nonkenge.* Retrieved October 17, 2022, from https://pod.link/epokothweni

Odine, M. (2013). Use of Radio to Promote Culture in South Africa. *Journal of Radio & Audio Media, 20*(1), 181–196.

Onyenankeya, K., & Salawu, A. (2019). Indigenous Language Radio, Identity, and Belonging. *International Journal of Diverse Identities, 19*(2), 33–49.

PwC. (2022). *PwC Africa Entertainment & Media Outlook 2022–2026.* PwC. Retrieved October 18, 2022, from https://www.pwc.com/ke/en/publications/entertainment-and-media-outlook.html

Setiowati, E. (2018). The Pattern Program of the Morning Radio Sequence: Study of Four Major Stations in Jakarta. *KnE Social Sciences*, 285–292. Retrieved December 7, 2022, from https://www.knepublishing.com/index.php/KnE-Social/article/view/2767/5988

Swiatek, L. (2018). The Podcast as an Intimate Bridging Medium. In *Podcasting: New Aural Cultures and Digital Media*. Springer International Publishing, (pp. 173–187).

Van der Merwe, R. L. (2021). Africa on Demand. In W. Mano & V. C. Milton (Eds.), *Routledge Handbook of African Media and Communication Studies*. Routledge.

CHAPTER 10

# Evolving Trends in Radio Broadcasting in South Africa: The Case of *Podcast and Chill with MacG*

*Gilbert Motsaathebe*

## Introduction

Podcasting is a form of radio that utilises digital sound that can be transmitted through various devices. The name podcast derives from the fact that it uses the technology developed by Apple in the form of the iPod. Thus, the word podcast is a combination of the words iPod and broadcast. According to Singer (2019, p. 572), the first "documented podcast was in 2004 when former MTV VJ Adam Curry used RSS feeds to push MP3 files to subscribers of his online show". In South Africa, podcasts have become a significant aspect of the radio landscape, playing a vital role in conveying different types of content according to each podcast's specific niche area. For example, podcasts such as *New Frame* [This podcast closed while this book was being produced] focus on current affairs and hard-hitting investigative stories and use audio podcast episodes.

G. Motsaathebe (✉)
ILMA, North-West University, Mahikeng, South Africa



S. Nkoala, G. Motsaathebe (eds.), *100 Years of Radio in South Africa, Volume 2*, https://doi.org/10.1007/978-3-031-40706-2_10

In contrast, others, such as *Podcast and Chill with MacG*, focus on light-hearted news and controversial stories mostly involving celebrities using video podcast episodes. As Mkhabela (2021, para 2) puts it, "The presence of podcasts in South African pop culture has been more distinctly felt…They have become crucial in documenting and dissecting culture and societal issues". There are several podcasts in the country, as shown in Table 10.1. They include *The Sobering*, hosted by Kitso Moremi, Kabelo Moremi and Javas Skolo, "one of the longest-running music and subculture podcasts in South Africa, having started in 2015" (Mkhabela, 2021, 02).

This chapter begins by providing an overview of *Podcast and Chill with MacG* and its significance as the case study considered in the research on which this chapter based. It then discusses why the frameworks of uses and gratifications theory and technological determinism theory have been used to evaluate podcasts in this context. The study employs content analysis as its analysis method and, in so doing, considers aspects such as the podcast's tone, mise-en-scène and the presence/lack of female representation. It then analyses the comments from listeners based on how the podcast gratifies their need for humour, natural conversation and off-the-cuff remarks that some view as controversial.

**Table 10.1** Some of the popular podcasts in South Africa in the past 10 years

| *Podcast name* | *Date started* | *Main host(s)* |
|---|---|---|
| *Podcast and Chill with MacG* | 2018 | Macgyver "MacG" Mukwevho, Sol Phenduka and Ghost Lady |
| *The Sobering* | 2015 | Kitso Moremi; Kabelo Moremi, aka Lil Frat; and Javas Skolo |
| *New Frame* | 2019–2022 (ended in 2022) | Charles Leonard, Musawenkosi Cabe, Bonolo Mokua, Tebadi Mmotla |
| *History of SA* | 2020 | Des Latham |
| *True Crime South Africa* | 2019 (June 22) | Nicole Engelbrecht |
| *African Music Concepts (Durban Gqom)* | 2016 | Ben Myster |
| *The Matt Brown Show* | 2015 | Matt Brown Show |
| *The Almost Perfect Podcast* | 2018 | Bop Perfect |
| *Frankly Speaking* | 2022 | Andrew Levy and Rorisang Tshabalala |
| *Gareth Cliff Show* | 2014 (May 1) | Gareth Cliff |

Source: Author

## A Brief Overview of *Podcast and Chill with MacG*

Dubbed South Africa's number one podcast, *Podcast and Chill with MacG* takes the form of a conversational interview with guests and sometimes just a conversation among the three presenters on a particular topic. The podcast broadcasts on a YouTube channel with a simulcast on Channel O of M-Net. It is hosted by Macgyver Mukwevho, who goes by the nickname MacG, which is also the eponymous name of the podcast series. He co-hosts with two other presenters who are equally zealous and strong-willed in their approach, Sol Phenduka and the enigmatic Ghost Lady, who never shows her face on camera. It broadcasts on Tuesdays and Thursdays. The podcast invites South African celebrities and interviews them on controversial topics involving their personal and professional lives.

Some of those previously featured on the podcast series include world-renowned disk jockey Black Coffee (Nkosinathi Maphumulo), musician and DJ Dr Malinga (Goodwill Malinga), actress and television producer Sonia Mbele, popular media personality DJ Sbu (Sbusiso Leope), actor Presley Chweneyagae, radio and television presenter Sizwe Dhlomo, actress Tebogo Thobejane, musician and *Uyajola 99* (reality television show) host Molemo Maarohanye also known as Jub Jub, media personality Pearl Thusi, socialite and entertainer Zodwa Wabantu, *Skeem Saam* (television soap opera) actor Clement Maosa, and musician and presenter Zola 7 (Bonginkosi Dlamini). According to MacG, "the celebrity episodes help bring attention to the podcast" (see Mkhabela, 2021, para 12). In addition to the crowd-pulling power of these stars, MacG also attributes the podcast's success to consistency, hard work and passion (Mkhabela, 2021).

However, the hosts' use of a no-holds-barred approach often lands the show in hot water. For example, it was sued by actress Amanda Du Pont after an explosive interview with Maarohanye, who revealed intimate details on air about his affair with Du Point. Du Point criticised MacG for dragging her name into the discussion "for clout at the cost of amplifying a gender-based violence case". She also posted an Instagram statement accusing him of rape and appealed to other women who may have been victims of gender-based violence at the hands of either Jub Jub or MacG to come forward. Three women responded by accusing Maarohanye of rape. They included another celebrity, former Metro FM presenter Masechaba Khumalo. As a result of the rape allegations, Moja Love, a television channel that hosts the popular reality television show *Uyajola99*, which Maarohanye presents, suspended him. Another controversy that landed the podcast in hot water relates to the remarks on transgender people that the podcast series' presenters MacG (Macgyver Mukwevho) and Sol Phenduka made on one of their episodes. As a result of the remarks, beer company Amstel withdrew its sponsorship of the podcast.

The presenters also tell exciting jokes, making the podcast series a darling of its audiences who log into the popular podcast for laughter and unfiltered content. These hosts are comedians in their own right. As indicated, the podcast series broadcasts on YouTube with a broadcast simulcast on M-Net's Television Channel O,[1] effectively bridging the gap between radio and television because of its televisuality. The mode and style of the podcast are egalitarian and tackle interesting issues involving South African celebrities. As Mkhabela (2021, para 03) says, "Owing to their long-form nature, interview-and conversation-based podcasts, which are prevalent in South Africa, are in-depth, uncensored, unfiltered and not restricted in length by traditional media time slots". The interview guests are celebrities who are quizzed on areas involving their general experience and personal and professional life. Another significant feature of the show is that the hosts are bubbly, brazen and audacious and speak their minds on any issue that is up for discussion. The podcast series has no distinctive signature tune but uses a popular tagline, *Hee ndaa…bare mene?* a Tshivenda phrase meaning "Greetings…how are you?"

The target audience of this podcast are young people who are into celebrity gossip and humorous content. As Edward Tsumele (2021) says, the language used in the show will make older and conservative audiences uncomfortable. This source maintains that the podcast's "core audience seems to be those that are young and are hungry for celebrity gossip, raw politics and raw sex talk, delivered in the language the youth do understand and often use in the streets with liberal abandon" (Tsumele, 2021, para 2). The fact that podcasts are not regulated in the same way as traditional radio allows more freedom for their hosts to use language more liberally and tackle topics that may be regulated on traditional radio, which tend to be more attractive to youth than the older generation. The podcast series has a huge number of subscribers and millions of viewers. It recently hosted an event to celebrate reaching one million YouTube subscribers (Adekunle, 2023).

[1] M-Net is South Africa's major pay television channel.

## Theoretical Framework

I find the uses and gratifications theory (UGT) developed by Blumler and Katz germane to the issues raised in this chapter. This is because the chapter considers how the audience engages with the podcast series and how they seem to derive pleasure from it. This is in line with the central tenants of the UGT, which asserts that individuals use the media to satisfy a particular need. It explores how people deliberately choose the media to fulfil their needs, for example, entertainment, education or escape from reality. Essentially UGT, as espoused by Blumler and Katz (1974), postulates that the audiences use media for the following: (a) Diversion or Entertainment, which implies escaping from reality; (b) Personal Relationships in which they substitute actual relationships with others; (c) Personal Identity in which they learn and unlearn behaviour based on the content and characters portrayed in the media; and (d) Surveillance in which the media serves to satisfy their need for information. New media has opened up new avenues of uses and gratifications, and *Podcast and Chill with MacG*, the subject of this study, is an example of such media.

Another theory that I find appropriate is the technological determinism theory, which in its simplest form implies that technology routinely creates new avenues which impact different facets of humanity, including culture. It holds that as technology continues to evolve, it is adopted, and in return, that adoption breeds new avenues of further technological developments and needs, resulting in relentless technological development which has become an essential aspect of humanity. Sociologist Thorstein Veblen first developed this theory before it was expanded by media scholars such as Marshall McLuhan. Its fundamental assumption is that technology leads to changes and new development, including new tools which become part and parcel of society.

The evolution of podcasts is seen as one of those evolving trends necessitated by technology. This theory has received considerable attention over the years and was also appropriated by Karl Marx to explain how technological changes influence social relations. In this chapter, I take this technological determinism perspective further to explain how changes in technology influence the consumption of media content and changing media consumption/production experiences in a way that is equally continual, empowering and daring. I do this against 100 years of radio in the South African context, where podcasts present a significant milestone on the audio content continuum. This theory is therefore deemed appropriate to make sense of issues relating to the development of this critical innovation (podcasting) and related questions raised in this chapter. It helps contextualise the development of this new radio form within the broader cultural and economic environment which clarifies trends, content, form and users' experience that characterise it.

## Methodology

This chapter looks at several aspects of *Podcast and Chill with MacG* through an analysis of select episodes of the podcast series, which is one of the most popular in South Africa. Examining the content will help clarify why the podcast thrives and resonates with its audience. Aspects studied include the audience's comments to gauge their reception of the podcast and delineate thematic areas of issues that the podcast's content privileged. The chapter also examines the visual and communicative aspects of the show. The latter will be achieved through content analysis, an appropriate analytical tool for graphical images, while the audience's comments will be studied through thematic analysis. The views from the audience provide a clue of what the audience loves about the show. The data from the users' comments were then presented thematically, drawing on the recurring themes from these comments. Similarly, the issues that became salient in the content analysis were arranged thematically.

As Berelson (1952, 01) stated, content analysis is "a research technique for the objective, systematic and quantitative description of the manifest content of communication". Notably, content analysis goes beyond overt expressions in the data to encapsulate the covert expressions that are not explicit but implied in the data. As a reliable systematic method of communication content research, content analysis has been used widely to analyse communication artefacts, including texts, audio and visuals (see Berelson, 1952; Krippendorff, 1980; Motsaathebe, 2009, 2012, 2018; Mathe & Caldwell, 2017). As indicated, the chapter focuses on the popular *Podcast and Chill with MacG.* The study used the following content analysis process: (1) selecting the episodes, (2) studying the episodes and the subsequent audience comments inductively to construct themes/categories for analysis, (3) analysis and (4) discussion and reporting of findings.

### *Selection of the Episodes for Analysis*

The study focused on episodes that aired between April 2022 and April 2023. The study employs a qualitative approach using qualitative content analysis. In terms of design, the chapter follows an exploratory approach which aims to chart ways for more detailed studies on this embryonic radio trend of podcasting. Six episodes were randomly selected. All the episodes are available online as the podcast series is broadcast on a YouTube channel. This sample size is considered sufficient for this study because, in qualitative research, the researcher is interested in rich data that will be analysed for insights, unlike in qualitative research, which prioritises a large sample.

## Data Analysis

A content analysis of the episodes from *Podcast and Chill with MacG* was undertaken to examine the topics they cover and their presentation style. As the show is broadcast online and the episodes are also available online, the analysis includes an examination of audience responses to see how they interact with both the presenters and the content broadcast. I first watched the episodes to delineate important issues that became the analysis categories of the study. This was in line with the demands of content analysis, which requires establishing categories for analysis. The analysed episodes are presented in Table 10.2:

Table 10.2 shows the episodes analysed for the study that informs this chapter. There were many episodes, but this study focused only on six to

**Table 10.2** Taxonomy of selected episodes

| |
|---|
| Episode 310 \| Black Twitter, Amanda Du Pont, Minnie Dlamini, Cancel Culture, Chillers Market |
| Episode 278 \| Only Fans, Zakes Bantwini, Boity, Home Loans, Somizi, All You Need is Neo |
| EPISODE 407 \| Sizwe Dhlomo, Radio, ANC, Glencore, Kabza De Small, Ye, Black Panther, Take Off, Davido |
| EPISODE 388 I Innocent Masuku on Yizo Yizo and Drug Abuse, Rehab, Reality Show |
| Episode 363\| Clement Maosa on Death of Parents, Witchcraft, *Skeem Saam*, King Monada, Patrick Shai |
| \|Episode 305\| House Managers, Amapiano Diva, Vuyo Dabula, Unathi Nkayi, SK Khoza, Tebogo Thobejane |

Source: Author

paint a broader picture of the issues covered, how they are covered, the style of presentation used by the show's presenters, and how the audience engages with the content and the presenters. First and foremost, the analysis perused the selected episodes to get an idea of the issues tackled in the podcast series. This kind of approach is called inductive because the researcher approaches the content with an open mind so that the categories of analysis emerge from the data, unlike in a deductive approach in which the researcher approaches the content with predetermined categories to investigate their presence or absence, prevalence and modality. This initial perusal revealed that the topics covered are wide-ranging but that they revolve around the following categories (Table 10.3):

### *Tone*

Furthermore, the perusal also looked at the tone used by the hosts as this is indicative of their style, which has an implication for the way the audience perceives them and how they relate to the guests of the show. The tone is, therefore, critical for the show's success. *Podcast and Chill with MacG* has a very distinctive tone which is casual, cheerful and humorous and has a sense of authenticity to it, a style that one viewer, Thato Motake, characterised as "pure and raw". All these attributes convey the exuberance of the hosts and play an important role in the amusement and how the audience generally experiences the show. As one audience member, Nkosinathi, puts it, "This podcast is unfiltered, and the chemistry between the hosts is phenomenal". There is always something to laugh about in almost every sentence uttered by the hosts.

**Table 10.3** Content categories (issues covered by *Podcast and Chill with MacG*)

| *Issues covered (categories)* | *Examples* |
|---|---|
| 1. Celebrity gossip | 1. Jub Jub about his jail time, girlfriends and use of drugs |
| 2. Personality profile interviews | 2. DJ Malinga losing his house |
| 3. LGBT issues | 3. Clement from *Skeem Saam* and SABC presenter |
| 4. Music, fame and money | 4. Home loans |
| 5. Hardcore, serious conversation: | 5. LGBT issues, witchcraft |

Source: Author

### *Mise-en-Scène*

Another aspect that is crucial in a content analysis study of this nature is mise-en-scène, which entails all props that appear on the screen. This is significant because this podcast series is televisual, meaning that it is an audio-visual programme that follows particular television aesthetics that all impact how the audience experiences it. The podcast is often broadcast from different locations. These locations include hotel pools, eateries and entertainment places. There is nothing distinctive about the attire of the hosts as they do not wear over-the-top clothing but stick to what could be characterised as streetwear. The advertisers' merchandise is often placed as part of the props on the set. For instance, in some of the episodes where the podcast was still sponsored by the beer company Amstel, bottles of Amstel were displayed on the set. How the hosts interact is also very interesting, as the interaction is conversational, witty and thought-provoking with a bit of hilariousness. The audience relates very well to this humorous approach to the conversation, as illustrated by how they partake in the dialogue in the comments in the chat section. The tagline—*Hee nda! Bare mene*—used by the main host at the beginning of the show could be considered as the main element of the mise-en-scène. It functions like the show's signature tune, reminding the audience that they are now watching this podcast series.

### *Female Voice(s)*

The media often privileges the voices and perspectives of men at the expense of women's voices and perspectives. The views of the enigmatic Ghost Lady, who is 'heard but not seen' on the show, fill this critical gap by ensuring that women's voices are heard on this important media platform. It is not clear why she is not appearing on screen while her two male co-presenters do. However, from a feminist perspective, the fact that she is not visible on the screen while her two male co-hosts appear could be seen as an invisibilisation. From this perspective, it means that she is being rendered invisible, compared to her male co-hosts, whom we see on screen all the time.

Invisibilisation is essentially the process of marginalisation. Motsaathebe (2012) found that the preponderance of men and the authoritative tenor that the media ascribes to them at the expense of female voices contributes to the production and perpetuation of gender stereotyping and imbalance in society. An invisibilisation of this kind further marginalises women. This is critical because the media are instrumental in cultivating gender stereotypes and play a pivotal role in undoing the stereotypes and promoting gender equality.

## Discussion

As indicated, the study focused on the evolving trend in radio with a focus on podcasting content and style of presentation. The role of presenters and the type of content of this podcast were seen as some of the factors that make it successful. This section presents an analysis of comments by the audience of the podcast series. Thus, the unit of analysis was the comments of the audience. Perusing the audience feedback in the form of comment threads on the YouTube channel on which the show is broadcast provides a good idea of how they feel about both the content and the presenters and their distinctive approach. Hence, the views of the audience were deemed crucial. The comments were analysed and categorised according to the themes around which issues cluster. Where comments are cited, they are quoted verbatim to retain their authenticity and reflect new media trends of using other communication codes, such as emojis, that also function semiotically to express emotions and help convey the intended meaning. Below are the themes that frame the gist of the comments.

### *Progressiveness*

Audience comments revealed that the topics covered in the show were cutting-edge and that the presenters made the show very interesting through their style of presentation, including being cocky. They also appreciated the show's progress and growth. For instance, after it was announced in one of the analysed episodes that the podcast series was partnering with M-Net and that it was going to be available on its Channel O, audiences reacted as follows:

> The audience appreciated the announcement that the show would be part of DSTV. @DJ Tiyani Sithole Channel.
>
> As Chillers we always knew we will be on DSTV most definitely this is the best podcast in the world of course Chiller. @I Am

What an emotional moment!! 😥😥😥. @Lucky Rankudu

Junior Kanye the prophet predicted it all congrats MacG and the team we definitely tuning in on Wednesday 21h30 yall know the vibes ✌❤🔥🔥🔥

As seen in the above comments audiences are able to interact with the show through their online comments. These sentiments generally reveal what the audience likes or dislikes about the podcast series. Some audiences are very elaborate in their chat, while others are very concise. These chats generally provide honest feedback on why they are watching the podcast series. This feedback from the audience could prove beneficial for the show's success, as it highlights areas that may require improvements.

### *Thriving on Controversy*

The podcast series often tackles controversial topics. The subjects discussed are usually topical issues that involve high-profile people and contentious aspects of their lives. For this reason, one audience member (Simphiwe Maseko) sees the podcast as a "one of a kind podcast". Others had this to say about the content of the podcast:

@Batandwa Dlali
The things you guys talk about in the podcast are the reason that only fans approached you.

@tomraynbp
I was just feeling down till this podcast popped up! Cheered up my mood.

*@V Ruben Shishaki 2*
*LOVE FROM NAMIBIA. YOU GUYS GOT ME OUT OF DEPRESSION. FOREVER GRATEFUL. CHILLER FOR LIFE.*

What the comments illustrate generally is that a successful media programme depends on multiple factors, including the content that the audience likes, the manner in which the content is presented, the presenter conveying the content, and how they interact with the audience and guests invited to participate in the show. For its part, the podcast series tackles controversial issues, and the presenters speak their minds without worrying about the consequences, and this seems to be something that the audience enjoys. The presenters seem aware of this and insist they do all that for their audience, as illustrated in the following comments by one of the podcast's hosts:

> What we do is for the chillers and whenever sh*t happens we get a lot of new chillers. So, we always benefit more than we lose from the shirt that happens.

In agreement with the above sentiments, one follower noted,

> @BibOw – THIS PODCAST IS EVERYTHING

Another by the name @Gladwin Letlojane said:

> I work a demanding Job at the Mine and am always under pressure. But I can assure you, I take 1hr of my day on the daily and the podcast really takes my mind off things. Keep up the good job guys. I love how well-researched it seems to be. I am currently in social work school so being able to listen to this even before I went back to school and now keeps me current on topics that seem reputable.

These comments suggest that the followers are die-hard fans of the show. Their strong sentiments dovetail with technological determinism's assumption that development brought by technology follows a predictable pattern in how society embraces it as it develops further. Similarly, the fact that the audience attests to their needs being met by the podcast series resonates with the uses and gratifications theory's assertion that people use media to gratify their needs and that they will leave a specific media channel if it does not meet their needs and will stay with those that gratify them.

### *Keep It Natural*

The phrase "keep it natural" refers to the overarching sentiments expressed by the audience with the way the show's presenters seem to rely firmly on their innate natural ability in their engagement on the show without following any discernible formulaic script. They are also not pretentious in how they engage with issues they discuss as they speak openly about how they feel without trying to pander to popular sentiments or to avoid any possible controversy. As one comment indicated:

@Leo Danstgo: One thing I like about this Podcast, in general, is that they always keep it natural, they don't hide feelings on whom to speak and what not to speak about. Keep it up! Love from Namibia 🇳🇦

### *Humorous Nature of the Show*

Humour has a positive effect on people. According to Barbara Field (2021, 01), "Humour might be like a soothing balm or a light diversion. But humour is much more powerful than something that lulls or calms us down. It has been proven that humour is good for your health and psychological well-being. For example, it can relieve stress and defuse tension. Followers expressed their amusement at the hilarious nature of the show and how this makes them feel.

@Tiisetso Lucas
What kills me about Mac G, he likes playing dumb and his actually a genius, that's how he leads many celebrities to the corner they don't want to go.

@Skinie Toei
Is It weird that I prefer watching with no guests? You guys alone are such entertainment 🔥🔥🔥 Celebs make it all serious.

### *Sense of Ownership on the Part of the Audience*

A sense of ownership is a feeling of identifying strongly with something in such a way that you feel you are part of it and want to see it succeed. This is reflected in the audience's comments in which they boldly declare that they are the "chillers", meaning that they are designated followers of the podcast series. In this way, they share in the failures and successes of the podcast. It is clear from their comments that they feel that they are an integral part of the show:

@Royale De'noir
I said to my man, 'we're gonna be on channel o' he's like 'who's we?' me and the other chillers duh, the podcast is ours lol 🤣🤣🤣🤣

@Kgopolo Tolo
Its crazy so f****n proud of the team. I feel like this show is mine personally too fucken happy. Let's go big things are coming for sure

@Relo Sikwane
Yikes, this is so so huge. Congrats guys! We are SO SO proud!

@Endurance HoldOn
Looking back from 2018 when the podcast had less than 5K subscribers. Mac G and this team deserves all the success. Mac G you were patient, true and persevered. 🥰

@anelisiwe mbo
This feels so personal to me 🥺🔥🔥. You guys deserve this and more. #for the love of the podcast, Asbonge! 🔥🔥

The other one, Nicky Mthelebofu, said she celebrated as if they had just gotten promoted. "Yoh the gig for Channel O made me happy, I celebrated like I just got a promotion at work. In fact, I celebrated as if I was given shares".

### *Knows No Boundaries*

It was clear that many followers liked the show and praised the presenters and their style. Most importantly, the podcast series' audiences were not only confined to South Africa. Some of the regular commenters were from Namibia and Botswana. For instance, one Namibian named Tenzy Temw said: "Yoo! The love I have for this podcast is compared to none. All the way from Namibia 🖤😊"

## Conclusion

This exploratory study discussed trends and patterns of podcast usage in South Africa against the backdrop of 100 years of radio in the South African context, where podcasts present a significant milestone on the audio content continuum. It used a content analysis of six episodes of *Podcast and Chill with MacG* to study the style of presentation by the hosts, and the issues they tackle on the show, as well as the audience comments to evaluate how they feel about this podcast series. I delineated the issues that tend to be covered on the podcast series as revolving around celebrity gossip, personality profile interviews, LGBT issues, fame and money, and hardcore and serious conversation. Using the uses and gratifications theory and technological determinism as a framework, the discussion endeavours to make sense of issues relating to the development of this critical innovation and contextualise the development of this new radio form within the broader cultural and economic role of audio media in this context.

The study found that the comments indicate how the audiences engage with the content of the podcast series and its presenters. Their remarks were delineated into themes that capture the essence of their likes and dislikes about the podcast series. These include the humorous nature of the show, a sense of ownership, the style of presentation by the hosts, the fact that the podcast series' reach extends beyond South African borders, and the unfiltered/natural nature of the show. These aspects make the podcast series unique and appealing to its audiences, which contributes to its massive success. One of the audience members articulated this fondness as follows: "Yoo! The love I have for this podcast is compared to none".

Another pertinent observation is the cockiness of the presenters and their resolve to speak their minds without worrying about the consequences. As they say, they are doing all that for their audiences or the chillers, as they call them. There was also an indication that the audience was thrilled to be part of the show and even celebrated the podcast's success as if they were celebrating their own individual success. This again resonates with the theory which suggests that this technology becomes an integral part of members of society. The audience's sense of ownership and being part of the chillers can be seen in the comment below as they even share ideas on what to include in the show:

> @Sandiso Majubana
>
> There needs to be a crossover episode between the Lady's Podcast and this one. The chemistry just works!

> Toolz deMaestro
>
> Imagine 105 minutes and the PODCAST is at 18,840 views. What we are seeing is pure talent!! Nice one PODCAST & Chill network.

It was also interesting to see how the audience showered the hosts with daring nicknames. Consider, for instance, this comment from Martin Masika: "Huge big-ups to Sol the Punisher, Ghost Lady the Silent Killer, MacG the Vandalizer and the Crew that doesn't screw. To more great things". Others, such as Letuka Skosana, had this to say about the enigmatic Ghost Lady, who is considered a voice of reason on the show. "Mac

Gee please take care of our Ghost Lady. She's the motherly figure that brings sanity to the podcast".

This chapter shows that podcasting has become part and parcel of the media landscape in South Africa as it is growing and becoming very popular. This is illustrated by the number of podcasts that continue to evolve and thrive. The chapter found that the podcast studied has become very popular with audiences who continue to engage directly with the content and show presenters through the comment/chat section available on the YouTube channel on which the show is broadcast. The study concludes that this kind of audience participation enhances public deliberations on issues of critical importance to society. The chapter also finds that its broadcast on YouTube, with the audience being able to watch in real time presenters as they go about their deliberation on the show, bridges the gap between radio and television as podcasting of this nature employs all television aesthetics that previously distinguished television from radio.

The chapter also surmises that this particular podcast's popularity could be attributed to its cutting-edge content, the style of presentation, and the fact that it invites celebrities who already have a massive following as stars in their own right. In terms of what the audience value most about the podcast series, five thematic categories emerged: the controversial nature of the topics, the humorous nature of the show, a sense of ownership, progressiveness, and the unfiltered/natural essence of the show. It was also significant that the podcast's influence extends beyond the borders of South Africa, with one audience member from Namibia saying that the podcast has helped her heal from depression.

Furthermore, the findings indicate that the podcast series has become an integral part of its audience, who could not wait for the next episode. One audience member (Mziwonke Mfiki) commented, "I felt like my life stopped when you guys delayed". These findings are particularly significant for their illumination of the uptake of this radio form and how the audience experiences it. These findings also provide a window into some factors that account for the success of *Podcast and Chill with MacG*. They show how the new technologies provided by the 4IR influence the media disciplines and help bring about new media forms and media consumption habits in which audiences have more power to engage with the content they consume. Future studies could consider including key-informant interviews with the producers and presenters of this popular show to examine, among others, the intricacies of running a successful podcast in an era characterised by a fragmented audience. Other studies could also consider this kind of online media from ethical and regulatory perspectives as these media forms are less regulated and do not have to adhere to strict codes of conduct governing professional media that is regulated.

## References

Adekunle, J. (2023). Mac G's "Podcast and Chill" Nears 1 Million Subscribers. *Fakaza News*. https://fakazanews.com/2023/03/02/mac-gs-podcast-and-chill-nears-1-million-subscribers/

Berelson, B. (1952). *Content Analysis in Communication Research*. Free Press.

Blumler, J. G., & Katz, E. (1974). *The Uses of Mass Communications: Current Perspectives on Gratifications Research* (Sage Annual Reviews of Communication Research, vol. 3) (pp. 292–318). https://eric.ed.gov/?id=ED119208

Field, B. (2021). The Health Benefits of Humor and Laughter. https://www.verywellmind.com/health-benefits-of-humor-and-laughter-5101137

Krippendorff, K. (1980). *Content Analysis: An Introduction to its Methodology*. Sage.

Mathe, L., & Caldwell, M. (2017). A Content Analysis of Readers' Comments on Political and Economic News Articles from Zimbabwe Online Media. *KRE Journal of Communication, 8*(1), 56–68. https://doi.org/10.31901/24566586.2017/08.01.174

Mkhabela, S. (2021). Like, Subscribe and Comment: The Rise of Podcasting in South Africa. *Mail & Guardian*. https://mg.co.za/friday/2021-07-02-like-subscribe-and-comment-the-rise-of-podcasting-in-south-africa/

Motsaathebe, G. (2009). Gendered Roles, Images and Behavioural Patterns in the Soap Opera Generations. *Journal of African Media Studies, 1*(3), 429–448.

Motsaathebe, G. (2012). The Use of Females as Sources of Information in SABC Newscasts. *Delta Journal Abraka: The Journal of Communication and Media Research, 2*(2), 13–24.

Motsaathebe, G. (2018). South African Cinema and its Depiction of Race, Gender and Class: Portrayal of Black Women in Post-Apartheid South African Films. *Canadian Review of Comparative Literature, 45*(3), 23–44.

Singer, J. (2019). Podcasting as Social Scholarship: A Tool to Increase the Public Impact of Scholarship and Research. *Journal of the Society for Social Work and Research, 10*(4), 571–590.

Tsumele, E. (2021). Popular Youth Oriented YouTube Channel Podcast and Chill with MacG Goes Mainstream as Chanel O Beckons. *City Life Arts*. https://citylifearts.co.za/popular-youth-oriented-youtube-channel-podcast-and-chill-with-macg-goes-mainstream-as-chanel-o-beckons/

CHAPTER 11

# The Road Ahead: Community Radio and the Digital Milieu—Challenges and Opportunities for the Next Century of South Africa's Radio Landscape

*Gilbert Motsaathebe and Sisanda Nkoala*

## Final Reflections

This volume adds to a growing corpus of research on radio in South Africa. To our knowledge, this volume forms the first comprehensive work that tackles the issue of radio in South Africa focusing exclusively on community radio and the digital milieu, highlighting the undercurrents of community radio and some of the evolving trends necessitated by technology. As with the first volume, *100 Years of Radio in South Africa: South African Radio Stations and Broadcasters Then & Now*, the current volume consists of contributions from media scholars and practitioners. These

G. Motsaathebe (✉)
ILMA, North-West University, Mahikeng, South Africa

S. Nkoala
Cape Peninsula University of Technology, Cape Town, South Africa





S. Nkoala, G. Motsaathebe (eds.), *100 Years of Radio in South Africa, Volume 2*, https://doi.org/10.1007/978-3-031-40706-2_11

contributions cover multiple aspects ranging from the history of community radio in the country to the espousal of technologies that resulted in new forms of radio such as podcasts and various streaming technologies. Thus, the volume highlights key developments in community media and the evolution of online radios as well as some of the key guideposts they use to attract their audience. In the final analysis, the chapters represented in this volume seek to forecast the place and character of South African radio in the coming years, drawing on multiple theories and approaches.

Other critical aspects tackled in the volume include the significance of community radio, its agility and the use of languages, especially indigenous African languages, as a key aspect of the affordances of this tier of radio broadcasting. The significance of community radio as highlighted in this volume is also borne by the fact that new community radio stations are being launched almost every year in South Africa, in addition to the over 200 community radio stations that we already have in the country.

The Independent Communications Authority of South Africa (ICASA) continues to call for these media applications. For instance, at the time that this book was being written, ICASA had opened another round of applications which was finalised in February 2023.

Furthermore, it is significant that some of the chapters in this volume chronicle the history of community radio in South Africa from the time of what was referred to as the Freedom of the Airwaves campaign up until the first community radio station was established. This is an important contribution because any serious discussion on community radio in South Africa would not be complete without this important aspect, as radio scholar Robin Sewlal points out: "Community radio is the first tier of broadcasting (commercial broadcasting and public service broadcasting comprise the other two) that has opened the airwaves in South Africa" (Sewlal, 2014, p. 24).

The volume also espouses the vital role that community radio plays when coming to social inclusion and public awareness. This is an important role in a society that was extremely fragmented due to the double tragedy of colonialism and apartheid that characterised South Africa. It is clear from this volume that by increasingly using indigenous languages, community radio contributes to media diversity and also plays a crucial role in the preservation of such languages and the cultures of those language communities, which some people have argued may face extinction if not utilised in public discourse. The analyses and arguments presented in the volume further indicate that using social media platforms such as

WhatsApp voice notes not only promotes citizen deliberation but also endorses the use of indigenous language in the digital public sphere.

Perhaps another important issue relating to digital technology and community radio is the fact that this technology complicates the notion of community, especially the geographical aspect of it as digital platforms mean that these media now transcend geographical boundaries. This will have implications for content and programming in the future. These technologies blur the line between the audio character of radio as it used to be since some of these platforms have essentially become audio-visual as in the case of many podcasts that are popular in South Africa.

The volume will appeal to media practitioners, radio historians, academics and students interested in media studies, journalism, communication and cultural studies. This volume (volume 2) and volume 1 represent the first substantial texts celebrating 100 years of radio in South Africa.

## Reference

Sewlal, R. (2014). The role and rise of community radio in the development of democracy in South Africa. *Rhodes Journalism Review, 34*, 21–26. https://rjr.ru.ac.za/rjrpdf/rjr_no34/The_Role_and_rise_of_community_radio.pdf

# Index[1]



[1] Note: Page numbers followed by 'n' refer to notes.



S. Nkoala, G. Motsaathebe (eds.), *100 Years of Radio in South Africa, Volume 2*, https://doi.org/10.1007/978-3-031-40706-2

The manufacturer's authorised representative in the EU is Springer Nature Customer Service Centre GmbH, Europaplatz 3, 69115 Heidelberg, Germany. If you have any concerns regarding our products, please contact ProductSafety@springernature.com

Printed and bound by CPI Group (UK) Ltd, Croydon, CR0 4YY
30/09/2026
02225311-0002